CompTIA Data+: DAO-001 Certification Guide

Complete coverage of the new CompTIA Data + (DAO-001) exam to help you pass on the first attempt

Cameron Dodd

<packt>

BIRMINGHAM—MUMBAI

CompTIA Data+: DAO-001 Certification Guide

Publishing Product Manager: Heramb Bhavsar
Content Development Editor: Joseph Sunil
Technical Editor: Sweety Pagaria
Copy Editor: Safis Editing
Project Coordinator: Farheen Fathima
Proofreader: Safis Editing
Indexer: Sejal Dsilva
Production Designer: Shankar Kalbhor
Marketing Coordinator: Priyanka Mhatre and Nivedita Singh

First published: December 2022

Production reference: 1301122

Published by Packt Publishing Ltd.
Livery Place
35 Livery Street
Birmingham
B3 2PB, UK.

ISBN 978-1-80461-608-6

www.packt.com

This book is dedicated to everyone out there who is working hard to improve themselves, fighting for a better life. I hope this book helps guide you on your journey.

Contributors

About the author

Cameron Dodd is a data professional who specializes in instruction and writing clean, simple curricula that can be understood without an advanced degree. He has helped thousands of people around the world find their place in the field of data analytics. Recently, he wrote and taught the CompTIA Train-The-Trainer course for Data+, where he taught instructors how to teach the information covered in this book.

I would like to thank the great team at Packt, who made this possible. Heramb, Kirti, Nazia, Joseph, Priyanka, Sweety, and everyone working behind the scenes came together to make this book what it is today. Finally, I would like to thank my wife, without whom I could not have done this. You are my support and my inspiration, now and always.

About the reviewer

Devin Moya is Data Science mentor from Jupiter, Florida. Since 2018, he has developed curriculum at top-rated Data Science bootcamps and mentored 450+ aspiring data professionals in more than a dozen countries. He specializes in being a data generalist, helping people from different backgrounds transition into data-related roles. He has a genuine love for the cutting-edge and when not working or spending time with his wife and kids, he is researching and building applications that combine AI, art, and human creativity.

Renganathan Palanisamy is a Data and AI Specialist and an avid trainer with 20+ years' experience in both academic and technical field. Serves as a Director with skills in Microsoft, AWS, Alibaba Cloud and IBM based Data and AI solutions. Renganathan also serves as Microsoft MCT Regional Lead to provide guidance to MCTs. He is also a pioneering CompTIA Data+ Trainer for APAC region. His involvement as mentor in technical communities is well known.

Table of Contents

Part 1: Preparing Data

1

2

3

Collecting Data   31

4

Cleaning and Processing Data   51

5

Part 2: Analyzing Data

6

7

Measures of Central Tendency and Dispersion   109

8

Common Techniques in Descriptive Statistics   129

9

Hypothesis Testing   145

10

Introduction to Inferential Statistics   161

Part 3: Reporting Data

11

12

13

Common Visualizations   225

14

Data Governance   241

Part 4: Mock Exams

Preface

CompTIA Data+ is an early-career data analytics certification for professionals tasked with developing and promoting data-driven business decision-making. Collecting, analyzing, and reporting on data can drive priorities and lead business decision-making. CompTIA Data+ validates that certified professionals have the skills required to facilitate data-driven business decisions. With this guide, you'll understand how to prepare, analyze, and report data for better insights.

You will start with an introduction to the Data+ certification exam format and then quickly dive into preparing data. You will learn about collecting, cleaning, and processing the data, along with data wrangling and manipulation. Moving on, you will learn to analyze data, in which we will talk about types of analysis, common hypothesizing techniques, and statistical analyses. We then will move on to reporting the data. We will learn about the common types of report visualizations and data governance. Later, you will test all the knowledge you've gained throughout the book with the help of some mock tests.

By the end of this book, you'll have covered everything you need to pass the AWS DBS-C01 certification exam and have a handy, on-the-job desk reference guide.

Who this book is for

If you are a data analyst looking to get certified by the DAO-001 exam, this is the book for you. This CompTIA book is also ideal for those who need help entering the quickly growing field of data analytics and are seeking professional certifications.

What this book covers

Chapter 1, *Introduction to CompTIA Data+*, covers what topics the CompTIA Data+ certification exam will include, as well as how the exam will be administered.

Chapter 2, *Data Structures, Types, and Formats*, discusses how data is stored, covering high-level concepts of database structures down to specific file types that you will need to know about.

Chapter 3, *Collecting Data*, covers all the methods that are available for you to use for collecting data.

Chapter 4, *Cleaning and Processing Data*, covers various types of data, and how to clean and process them for use.

Chapter 5, *Data Wrangling and Manipulation*, teaches you how to shape the data into a form that various applications can use.

Chapter 6, Types of Analytics, covers the most common types of analyses, and various examples of each of them. You will learn to perform EDA, performance analysis, trend analysis, and various methods for each.

Chapter 7, Measures of Central Tendency and Dispersion, introduces you to measures of central tendency and measures of dispersion, as well as how to calculate them. You will learn what a probability distribution is, and understand dispersion, means, medians, and various other measures.

Chapter 8, Common Techniques in Descriptive Statistics, covers specific analytical techniques that are commonly used in data analytics and how to use them to calculate various statistics.

Chapter 9, Hypothesis Testing, looks at the fundamental concepts you need to understand about hypothesis testing and why it is important. You will learn about null and alternative hypotheses, the p-value, and other concepts.

Chapter 10, Introduction to Inferential Analysis, covers various types of statistical analyses, which will help you understand data better.

Chapter 11, Types of Reports, covers various types of reports, and how you can use each of them for various types of data analysis.

Chapter 12, Reporting Process, covers how you can create a report based on a question, or the structure and format that you need to follow for it.

Chapter 13, Common Visualizations, covers the most common visualizations that you can use to showcase your data, including heatmaps, treemaps, bubble charts, and many more.

Chapter 14, Data Governance, covers how you can manage your data and keep it secure. You will learn about various security measures and requirements, and you will understand what a data breach is and how to react to it.

Chapter 15, Data Quality and Management, covers various data control and quality checks, which will ensure that your data reports are accurate and useful.

Chapter 16 and Chapter 17, Practice Exams One and Two, will put you through a couple of practice exams that are designed around the Data+ exam so that you can prepare yourself for it. They have answers to every question, along with the context and explanation for each of them.

To get the most out of this book

Software/hardware covered in the book	Operating system requirements
Python 3.x	Windows, macOS, or Linux

If you are using the digital version of this book, we advise you to type the code yourself or access the code from the book's GitHub repository (a link is available in the next section). Doing so will help you avoid any potential errors related to the copying and pasting of code.

Download the example code files

You can download the example code files for this book from GitHub at `https://github.com/PacktPublishing/CompTIA-Data-DAO-001-Certification-Guide`. If there's an update to the code, it will be updated in the GitHub repository.

We also have other code bundles from our rich catalog of books and videos available at `https://github.com/PacktPublishing/`. Check them out!

Conventions used

There are a number of text conventions used throughout this book.

`Code in text`: Indicates code words in text, database table names, folder names, filenames, file extensions, pathnames, dummy URLs, user input, and Twitter handles. Here is an example: "If you would like to follow along, you can find the `EDA_Example_Data.csv` dataset by following the link provided."

A block of code is set as follows:

```
import pandas as pd
MyData = pd.read_csv("EDA_Example_Data.csv")
MyData.head()
```

Bold: Indicates a new term, an important word, or words that you see onscreen. For instance, words in menus or dialog boxes appear in **bold**. Here is an example: "Select **System info** from the **Administration** panel."

> **Tips or important notes**
> Appear like this.

Get in touch

Feedback from our readers is always welcome.

General feedback: If you have questions about any aspect of this book, email us at customercare@packtpub.com and mention the book title in the subject of your message.

Errata: Although we have taken every care to ensure the accuracy of our content, mistakes do happen. If you have found a mistake in this book, we would be grateful if you would report this to us. Please visit www.packtpub.com/support/errata and fill in the form.

Piracy: If you come across any illegal copies of our works in any form on the internet, we would be grateful if you would provide us with the location address or website name. Please contact us at copyright@packt.com with a link to the material.

If you are interested in becoming an author: If there is a topic that you have expertise in and you are interested in either writing or contributing to a book, please visit authors.packtpub.com.

Share Your Thoughts

Once you've read *CompTIA Data +: DAO-001 Certification Guide*, we'd love to hear your thoughts! Scan the QR code below to go straight to the Amazon review page for this book and share your feedback.

https://packt.link/r/1-804-61608-7

Your review is important to us and the tech community and will help us make sure we're delivering excellent quality content.

Download a free PDF copy of this book

Thanks for purchasing this book!

Do you like to read on the go but are unable to carry your print books everywhere? Is your eBook purchase not compatible with the device of your choice?

Don't worry, now with every Packt book you get a DRM-free PDF version of that book at no cost.

Read anywhere, any place, on any device. Search, copy, and paste code from your favorite technical books directly into your application.

The perks don't stop there, you can get exclusive access to discounts, newsletters, and great free content in your inbox daily

Follow these simple steps to get the benefits:

1. Scan the QR code or visit the link below

https://packt.link/free-ebook/9781804616086

2. Submit your proof of purchase

3. That's it! We'll send your free PDF and other benefits to your email directly

Part 1: Preparing Data

This part provides an overview of the CompTIA Data+ certification exam, before jumping right into the process of preparing data for analysis. Starting with how data is stored, you will walk through the process step by step for collecting, processing, and shaping your data into something you can use.

This part covers the following chapters:

- *Chapter 1, Introduction to CompTIA Data+*
- *Chapter 2, Data Structures, Types, and Formats*
- *Chapter 3, Collecting Data*
- *Chapter 4, Cleaning and Processing Data*
- *Chapter 5, Data Wrangling and Manipulation*

1
Introduction to CompTIA Data+

Welcome! This book is designed specifically to teach you everything you need to know to pass the CompTIA Data+ (DAO-001) certification exam. Here, you will learn the ins and outs of each domain covered by the exam, before going over practice questions at the end of every chapter to solidify those lessons. The last chapter of the book contains two mock exams that will test your knowledge and see whether there are any areas you should review before taking the exam. While CompTIA Data+ is a certification worth having, testing for it is not cheap, so we are going to do everything we can to empower you to pass on your first try. Ready? Let's begin.

This chapter is an introduction to the certification itself. Together, we will talk about what Data + is and the benefits of achieving it. Then, we will briefly go over what sorts of content will be covered by the exam. Finally, we will discuss the format of the exam itself, going over things such as how long you have to take the exam, how many questions there will be, and how they will be distributed across the different exam domains. Our goal here is to understand what Data+: DAO-001 is, so we can start preparing you for it.

In this chapter, we're going to cover the following main topics:

- Understanding Data+
- Introducing the exam domains
- Going through the exam format

Understanding Data+

Before we dive into specific domains covered by the test and how it is structured, we should briefly discuss what the Data+ certification is and why you should obtain it, as well as why data science is a good field to enter right now. It is important to have a clear understanding of your motivations before you begin.

CompTIA Data+: DAO-001

CompTIA is an organization known for professional and technical certifications, particularly in the field of cybersecurity. The certifications they provide indicate that a person has the knowledge and skill set to perform a specific task. Earning one of these certifications means that you have more credibility in the field and are more likely to be able to enter or move up in the field of your choice. **Data+: DAO-001** is the first certification CompTIA has created for data science. It is also one of the first data science certifications by any of the major professional certification companies. This is a major step in standardizing terminology and roles within the field.

Because data science is still a developing field, people are coming in from all sorts of educational and professional backgrounds. What this means is that employers are often not sure what specifically they need to look for and so certifications are becoming more important. As we have seen with some other technical fields, such as cybersecurity, it is difficult to get employed no matter what your background is, unless you have a certification from a recognized organization such as CompTIA.

> **Important note**
>
> Data+: DAO-001 is a vendor-neutral exam. Because data science is so varied, no one tool or software is used in every role. To reflect this, no specific programming language is required to pass this exam. That said, this book uses popular tools or programming languages as examples to help explain certain concepts. These examples are only one way to reach a solution and are meant to reflect broader concepts. You should practice using the tools with which you are most comfortable.

Okay, we've established that the CompTIA Data+: DAO-001 certification will help your data science career, but why should you enter the field of data science anyway?

Data science

Data science as a field is simply analyzing data to answer specific questions to help people make better decisions. It involves a little bit of statistics, a moderate amount of programming, and a lot of critical thinking. The concept has been around for a long time; the term **data science** has even been around since the 1960s. But it is only in the last few years that companies have gotten access to more data than they know what to do with. Large companies aren't the only ones with a lot of data anymore. Now, medium and even small companies are quickly gathering or purchasing large datasets. All these companies, large or small, now need professionals to store, access, analyze, and report on all this information.

Almost every list you find of the fastest-growing careers in the world will mention data scientist, or one of the other specialties in the data science domain, such as data analyst, data engineer, database administrator, research analyst, statistician, or half a dozen other job titles that end in **analyst**. It has been gaining in popularity this past decade and is only expected to continue growing in the next decade.

Data science is an exciting field that is quickly changing and developing, with new programs and techniques coming out every year. As the field continues to grow, there will be more and more jobs and opportunities for those who have the right skills and the right certifications. Now that we know why the test is important, let's talk about what this exam covers.

Introducing the exam domains

The exam was designed by a group of subject matter experts with different specialties in the field of data science. Together, they decided on common ground that any early career data analyst should know. They then categorized that knowledge into the following five domains:

- Data Concepts and Environments
- Data Mining
- Data Analysis
- Visualization
- Data Governance, Quality, and Control

Data Concepts and Environments

The domains move through the data pipeline chronologically. The first domain, **Data Concepts and Environments**, is largely about how data is stored. This covers multiple levels, from different database types, structures, and schemas, through file types for specific kinds of data, and even into different variable types. This domain is a broad view of storage concepts mixed with the ability to identify what type of data you can expect from different storage solutions.

Data Mining

This domain is a bit of a misnomer. **Data mining** is when you already have a huge dataset and you just go through it to find any insights that might be of interest, instead of answering specific questions. While data mining, you must go through all the concepts contained within this domain, but you also go through all these concepts for regular data analysis. What this domain is actually about is every step after storing your data but before you run an analysis. This domain includes collecting, querying, cleaning, and wrangling data. Effectively, these are the steps you need to take to get your data into a useful shape so you can analyze it.

Data Analysis

You have stored your data, you have pulled your data and made it pretty, and now it is time to do something with it. This domain is all about analyses. You will be expected to perform descriptive statistical analyses, understand the concepts behind inferential statistics, be able to pick appropriate types of analysis, and even know some common tools used in the field. You don't need to be able to use any of these tools because the test is vendor-neutral, just be able to identify them.

Visualization

It doesn't matter how perfect your analyses are if you can't communicate the results. What's the point in coming up with an equation that solves world hunger if you can't explain it to anyone else? To that end, the next domain is all about **visualizations and reporting**. This covers what information a report should include, what type of report is most appropriate, who should get a report, when reports should be delivered, the basics of report design, types of visualizations, and even the process of developing a dashboard.

Data Governance, Quality, and Control

The final domain is made up of larger concepts that span the entire life cycle of data analytics. A large part of this is made up of policies. Some of the policies focus on protected data and how it can be handled legally, while other policies are more about how you can ensure the quality of your data. If your data has low quality, you can't trust anything it says, and if you are mishandling protected information, you could face legal penalties, so these are important factors to know. This domain also includes a short section on the concept of master data management, as an example of an ideal state.

Now that you know what domains will be covered on the certification exam, let's talk about how the exam is structured.

Exam format

CompTIA Data+: DAO-001 was launched on February 28, 2022. You have 90 minutes to answer 90 questions. Most of the questions will be multiple choice, but there will be some performance-based questions, in which you will be presented with an example of something you would encounter on the job and you must make appropriate decisions based on it. Scoring is on a scale of 100 to 900, with 675 being considered a passing score. For the most up-to-date information on the exam, you can check out the exam website at `https://www.comptia.org/certifications/data`. The testing provider is Pearson VUE, and you can take the test online or at one of their approved testing centers. The following table breaks down roughly how many questions will fall into each domain:

Data Concepts and Environments	Data Mining	Data Analysis	Visualization	Data Governance, Quality, and Control
15%	25%	23%	23%	14%

Table 1.1 – Percentage breakdown of each domain

Who should take the exam?

The test is aimed at those early in their data analytics career – specifically, those who have been working in a data analytics position for 1.5 to 2 years. This is not a requirement but a suggestion, because people with less experience than this may not have encountered all the concepts covered by the exam. At the end of the day, anyone who is eager to get further in the exciting field of data science should take this exam. Rest assured that even if you don't have personal experience with any of the concepts on the exam, they are covered here in this book!

Summary

In this chapter, we learned that CompTIA Data+: DAO-001 is a new professional certification in the rapidly growing field of data science, which is the study of gaining insights from data to inform decisions. This certification will confirm your skills to current or potential employers, helping you progress in your career in this field. The exam gives you 90 minutes to answer 90 questions that cover the domains that we discussed in the *Introducing the exam domains* section of this chapter.

Most importantly, you should know that, with the help of this book, you can pass this exam and take an important first step into the big wide world of data science.

In the next chapter, we will dive right into the first domain and look at data structures, types, and formats.

2

Data Structures, Types, and Formats

This chapter is all about data storage. Occasionally, a data analyst will have to collect their own data, but for many, the company already has data stored and ready for use. During their career, a data analyst may encounter data stored in several different formats, and each format requires a different approach. While there are hundreds, if not thousands, of different formats, the exam only covers the most common, and this chapter will go over what you need to know about them.

Here, we will discuss things such as how data is actually stored in a database, which includes whether or not it is structured, deciding what kind of data is stored in it, and whether the database is organized to follow a specific data schema to make it more efficient. We will also cover common database archetypes, or data storage solutions that are arranged in a specific way for a specific purpose, such as a data warehouse or a data lake. After that, we will discuss why updating data that is already stored can be problematic and a few approaches you can take depending on your goals. Finally, we will wrap up this chapter by discussing data formats so that you can tell what kind of data to expect from each format.

In this chapter, we're going to cover the following main topics:

- Understanding structured and unstructured data
- Going through a data schema and its types
- Understanding the concept of warehouses and lakes
- Updating stored data
- Going through data types and file types

Understanding structured and unstructured data

More often than not, data is stored in a database. Databases are simple places where data can be electronically stored and accessed. They can be stored on a single computer, a cluster, or even in the cloud. Databases come in every shape and size, but all of them fall into one of two categories:

- Structured

- Unstructured

Structured databases

If a database is structured, it follows a standardized format that allows you to set rules as to what kind of data can be expected and where. That is to say that there is, hopefully, a clear and logical structure to how the data is organized. There are two main database archetypes that the exam considers structured:

- Defined rows/columns

- Key-value pairs

Defined rows and columns refer to tables or spreadsheets. This is the format that most people are familiar with and is, by far, the most common in the field of data analytics. Usually, every column in the grid represents a variable or type of data collected, and every row represents a single entry or data point. The following screenshot shows a simple table that counts beans:

Total Number of Beans	Red Beans	Blue Beans	Yellow Beans
10	3	4	3
12	2	6	4
11	3	5	3

Figure 2.1 – Structured database: defined rows and columns

In the preceding screenshot, the columns are vertical and represent the different variables, while the rows are horizontal and represent a different entry or data point. The cells are specific values or counts that make up a data point.

Key-value pairs store each data point as a data object; every object in a series has the same set of keys, but the values for them can be different. In the following code snippet, each key represents a variable, and each value is the data collected for it, while each object is a data point:

```
"Beans" : [
    {
        "Total" : 10,
```

```
            "Red"  :  3,
            "Blue"  :  4,
            "Yellow"  :  3
        },
        {

            "Total"  :  12,
            "Red"  :  2,
            "Blue"  :  6,
            "Yellow"  :  4
        },
        {

            "Total"  :  11,
            "Red"  :  3,
            "Blue"  :  5,
            "Yellow"  :  3

        }
    ]
```

This is the same information as shown in *Figure 2.1*, but now it is stored with each row being a data object defined by { }. The column names ("Total", "Red", "Blue", and "Yellow") are the keys. The keys are unique in that they cannot be repeated within a data object. For example, you cannot have two keys named "Yellow". However, all objects within a set usually have the same keys. The numbers represent the values or are the same as individual cells within a table.

Unstructured databases

Unstructured data has, more or less, no attempt at organization. You can think of these as big buckets of data, often in the form of a folder, where individual files or random data objects can be dropped. The exam breaks these down into two groups:

- Undefined fields
- Machine data

In this case, an undefined field is sort of a catch-all for file types that do not fit into a structured database nicely. These include several different file types, such as the following:

- Text files
- Audio files
- Video files

- Images

- Social media data

- Emails

These are all data types that, by default, store every data point as a separate file, which makes it difficult to keep them in a conventional structured database.

The other type of unstructured data recognized by the exam is machine data. When data is automatically generated by software without human intervention, it is often considered machine data. This includes automated logging programs used by websites, servers, and other applications, but it also includes sensor data. If a smart refrigerator logs measurements of temperature and electricity used at a regular time interval, this is considered to be machine data.

Relational and non-relational databases

Another way to categorize databases is whether they are relational or non-relational. Relational databases store information and how it relates to other pieces of information, while non-relational databases only store information.

This gets a little confusing when we look at query languages for pulling information out of databases. The majority of query languages are variations of **Structured Query Language (SQL)**. Anything that isn't based on SQL is considered NoSQL. Now, in common usage, everything that uses SQL is structured and relational, and everything that is NoSQL is unstructured and non-relational. However, this is not quite true. This can get a little confusing, so we will break it down into the following four main points:

- All SQL databases are structured and relational

- All non-relational databases are unstructured

- Some NoSQL databases are structured and relational

- Some NoSQL databases are unstructured, but still relational

Let's start from the top. *All SQL databases are structured and relational*. SQL databases are broken down into tables. Tables are inherently structured and show how one point connects to another, so they are relational. For example, if you look at a table, you know how every cell in a column is related (they are all different data points that record the same measurement or classification) and how every cell in a row is related (they are all different measurements of the same data point).

All non-relational databases are unstructured. This means that all databases that do not show how things are related inherently have no structure. Any structural organization would show relationships. For example, if you have a database that is nothing but a folder for audio files, there is nothing that shows the relationships between the files. Each audio file is its own discrete unit, and there is no structure between them.

Some NoSQL databases are structured and relational. Just because something is not based on SQL does not mean that it has no structure. For example, the exam considers key-value pairs to be structured and relational, but these are most often used in JSON files, which are considered NoSQL.

Some NoSQL databases are unstructured, but still relational. There exist databases called **graph databases**. These are NoSQL databases, where the nodes are stored separately in an unstructured manner, but they also store the relationships between every node. If a database stores the relationships between each node, it is relational.

This seems a little jumbled, but the most important things to remember are that tables and key-value pairs are structured and relational, while undefined fields and machine data are unstructured and, unless otherwise stated, non-relational. This can be summarized in *Figure 2.2*:

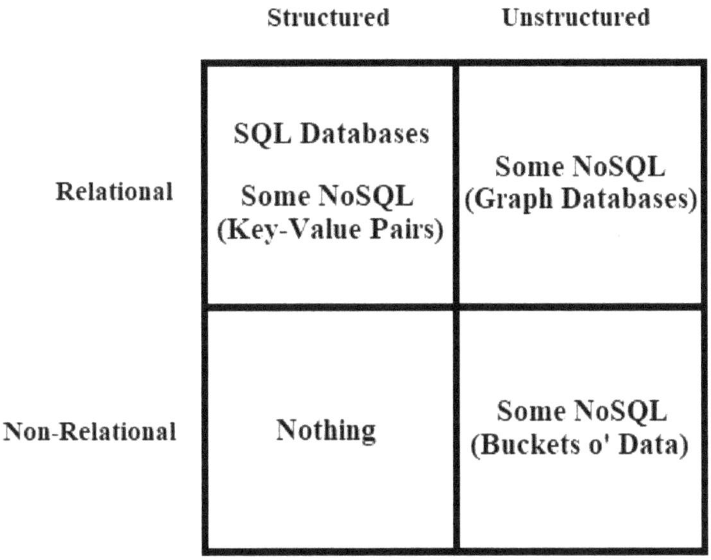

	Structured	Unstructured
Relational	**SQL Databases** **Some NoSQL (Key-Value Pairs)**	**Some NoSQL (Graph Databases)**
Non-Relational	**Nothing**	**Some NoSQL (Buckets o' Data)**

Figure 2.2 – Structure and relationships

Now, whether something is relational or structured is not the only consideration you need to think about when you are storing data in a database. How the tables themselves are arranged makes a huge difference in how useful and efficient a database is, so it's time to learn about schemas.

Going through a data schema and its types

An SQL database—structured and relational—is often made up of more than one table. In fact, more complicated databases may have dozens or even hundreds of tables. Every table should have a key. A key is a variable that is shared with another table so that the tables can be joined together. We will discuss specifics on joins later in this book. In this way, all the tables can be connected to one another, even if it would take several joins to do it. As you can imagine, these databases can be confusing, inefficient, and impractical. To make databases cleaner and easier to use, how tables are organized and interact with each other often follows a few common patterns. These patterns are called **data schemas**. There are several different popular schemas, and each meets specific needs, but this exam only covers two of the most basic schemas:

- Star
- Snowflake

The schemas are named after the shapes the tables make when you graph out how they are related.

Star schema

A **star schema** is one of the simplest schemas. At the center is a key table (sometimes called a fact table) that holds metrics, but they also have key variables for every other table in the database. Around the key tables, there are tables called **dimension tables**. Each dimension table has one key variable to connect to the key table and several other variables for storing information. Because all the dimension tables are connected directly to the central key table, the shape looks like a star. In *Figure 2.3*, you can see an example of the basic database of this schema, with the key table in the center and the dimension tables attached:

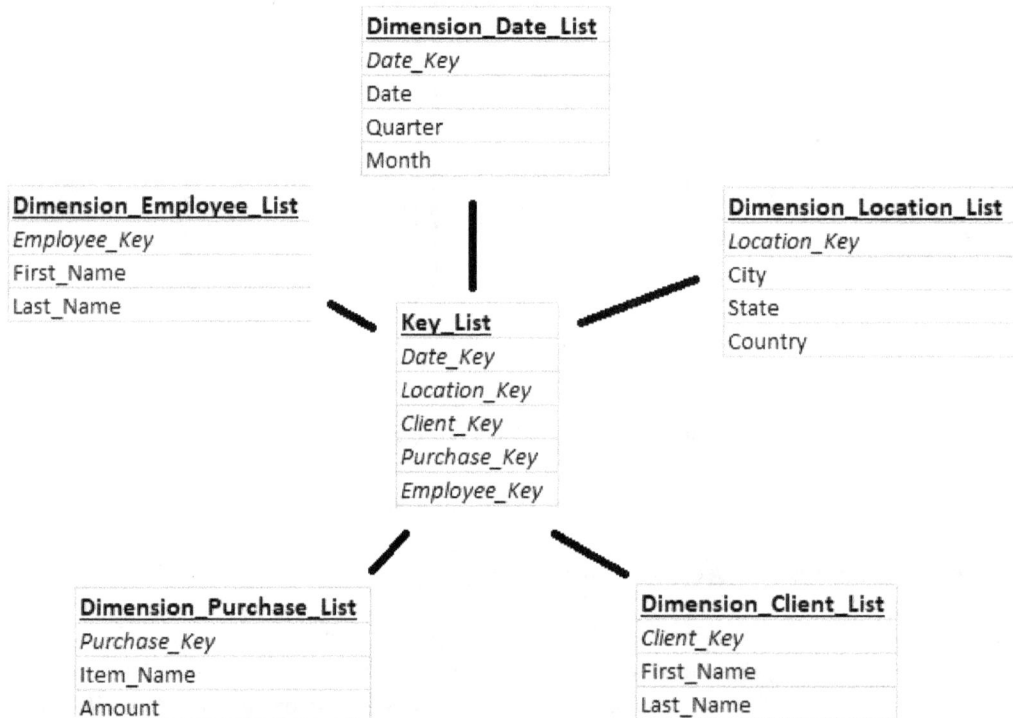

Dimension_Date_List

Date_Key
Date
Quarter
Month

Dimension_Employee_List

Employee_Key
First_Name
Last_Name

Dimension_Location_List

Location_Key
City
State
Country

Key_List

Date_Key
Location_Key
Client_Key
Purchase_Key
Employee_Key

Dimension_Purchase_List

Purchase_Key
Item_Name
Amount

Dimension_Client_List

Client_Key
First_Name
Last_Name

Figure 2.3 – Star schema

As you can imagine, there are pros and cons to this type of schema.

Pros:

- Simple (there are generally fewer tables)
- Fewer joins are required
- Easier to understand how the tables relate to one another

Cons:

- High redundancy (a lot of data is repeated)
- Denormalized (because of the high redundancy)

This schema is more user-friendly, but not the most efficient for larger databases. In *Figure 2.4*, you will see a diagram detailing a join with a star schema:

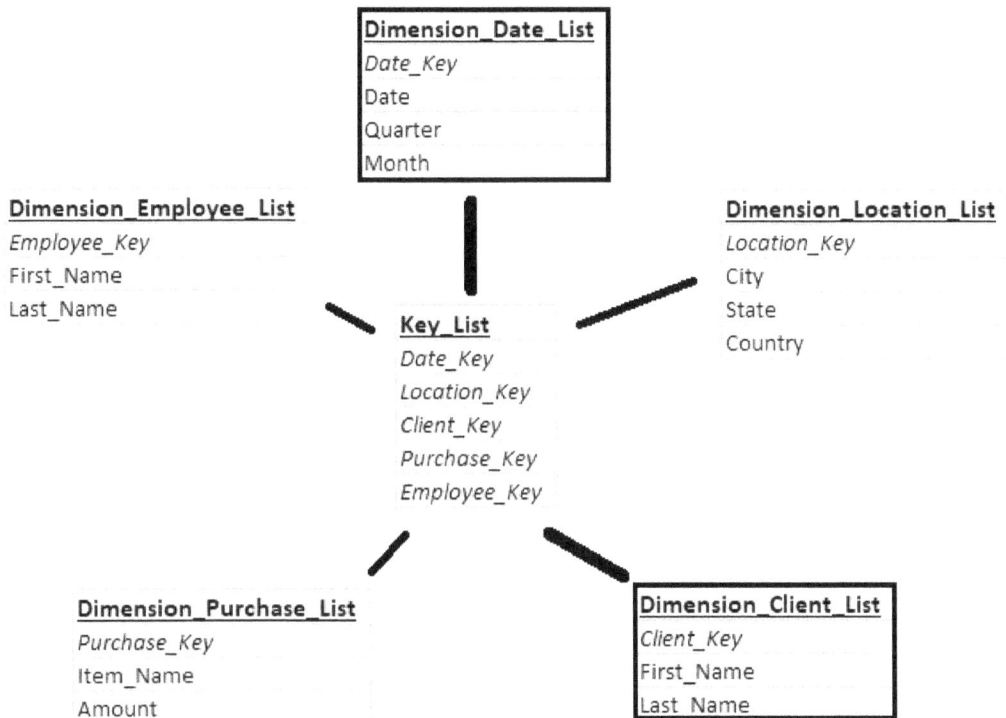

Dimension_Date_List
Date_Key
Date
Quarter
Month

Dimension_Employee_List
Employee_Key
First_Name
Last_Name

Key_List
Date_Key
Location_Key
Client_Key
Purchase_Key
Employee_Key

Dimension_Location_List
Location_Key
City
State
Country

Dimension_Purchase_List
Purchase_Key
Item_Name
Amount

Dimension_Client_List
Client_Key
First_Name
Last_Name

Figure 2.4 – Joining with a star schema

Imagine you wanted information from `Dimension_Date_List` and `Dimension_Client_List` together. You only need to join `Dimension_Date_List` to `Key_List` and then `Key_List` to `Dimension_Client_List`. That is only two joins, and you will never need more than two joins to connect any two tables in a star schema. That said, because the information is condensed to fit around a single table, information is often repeated, and it is not the most efficient approach. For example, in the preceding diagram, `Dimension_Date_List` has a date variable, but then it also has quarter and month variables in the same table. The variables are repeated because they are used for different things, but are grouped together in a table in a star schema.

Snowflake schema

A **snowflake schema** is similar to a star schema with one main difference: there are two levels of dimension tables. There is still a key table in the middle with dimension tables connected directly to it, but there is now a second set of dimension tables that connect to the first. This doesn't necessarily mean that there is more data than we saw in the star schema, but the data is spread out more. *Figure 2.5* is a simplified example of what a snowflake schema may look like:

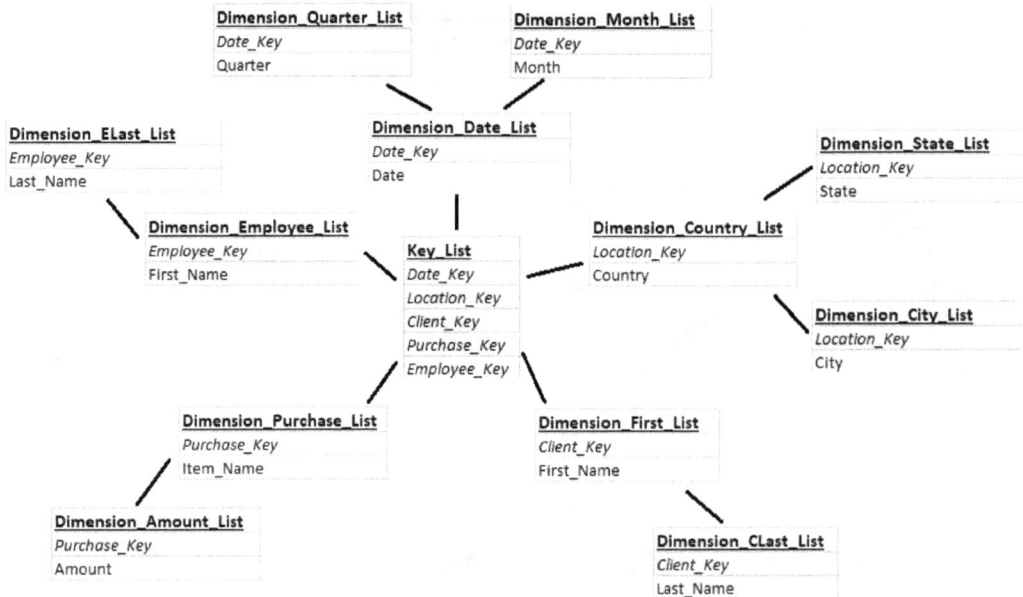

Figure 2.5 – Snowflake schema

Because of the branching lines coming out from the center, it is considered to look like a snowflake. Just as with a star schema, a snowflake schema has strengths and weaknesses.

Pros:

- Low redundancy (very few, if any, metrics are repeated)
- Normalized (follows a step of rules introduced by Edgar F. Codd to optimize databases)

Cons:

- More complicated (understanding how the tables relate to one another is more difficult)
- More joins are required (joining two tables may require several intermediate joins)

These are generally more efficient than star schemas, but they are less user-friendly. Because they are more complicated, it requires a greater understanding of how this specific database is structured in order to navigate it. *Figure 2.6* is an example of trying to join two random tables within a snowflake schema:

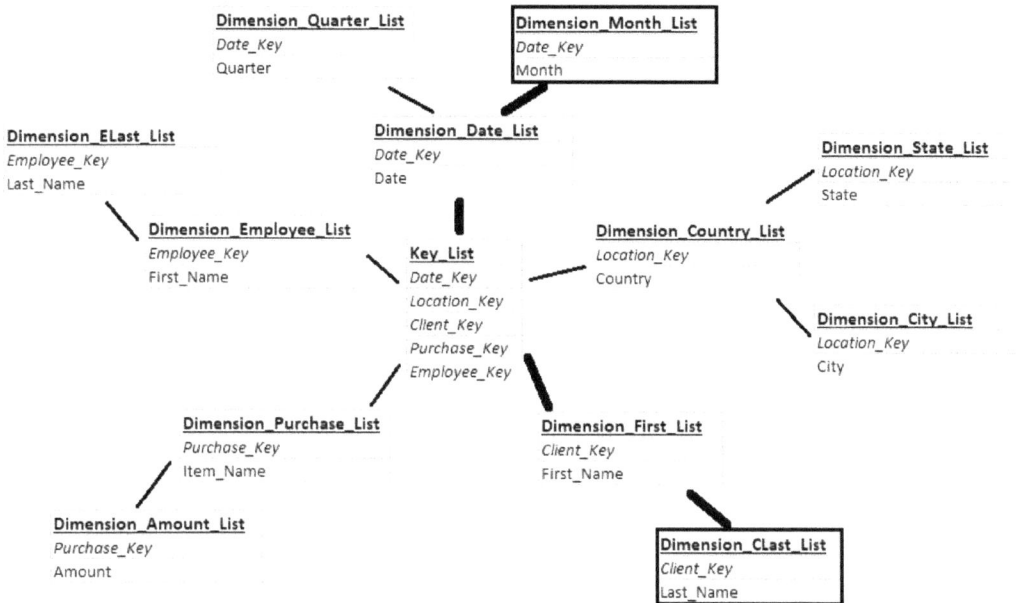

Figure 2.6 – Joining within a snowflake schema

Here, connecting any two tables may require anywhere from two to four joins. For example, if we wanted to connect `Dimension_Month_List` to `Dimension_CLast_List`, we would have to connect the `Dimension_Month_List` table to `Dimension_Date_List`, `Dimension_Date_List` to `Key_List`, `Key_List` to `Dimension_First_List`, and the `Dimension_First_List` table to `Dimension_CLast_List`. A snowflake schema only has one more level of tables than a star schema, but you can probably see how it can already be more complicated. However, the data is more spread out and does not repeat itself as much, making it more efficient. In the preceding simplified example, we see that `Date` has now been spread out among three tables, so there is no repeated information in any one of them.

Next, we will talk about a way to classify databases based on how they are used.

Understanding the concept of warehouses and lakes

Not all databases are used for the same purpose—they often become specialized based on how they are used. Each specialized database has a specific name, is used for different things, has different kinds of data, and is used by different people. A few of these specialized databases include:

- Data warehouses
- Data marts
- Data lakes

There are more, but these are some common types of database archetypes that you are likely to encounter. More to the point, these are the ones that will be covered in the exam.

Data warehouses

Data warehouses are most often used for structured relational tables. Usually, they hold large amounts of processed transactional data. Data warehouses are often more complicated and are used by data engineers or database administrators. Because these store large amounts of data and efficiency is more important, they are more likely to follow a snowflake schema.

Data marts

Data marts are a specialized subset of a data warehouse. They are smaller, only hold processed information on a specific topic, and have a simpler structure. Data marts usually contain customer-facing data and are considered self-service because they are designed to be simple enough for analysts or customer support employees to access by themselves. Because these databases prioritize ease of use, they often follow a star schema.

Data lakes

Data lakes store large amounts of raw, unprocessed data. They can contain structured data, unstructured data, or some combination of both. Data lakes often collect and pool data from several different sources and can include different data or file types. Because of the nature of raw information, data lakes are often used by data scientists and do not follow any specific schema.

> **Important note**
>
> At this stage in your career, you will not be asked to create a data warehouse or lake. These are usually made by very specialized data engineers, and many companies only have one. That said, the majority of data warehouses and lakes are created through third-party services or software. Some of the more popular approaches include Snowflake, Hevo, **Amazon Web Services** (**AWS**) data warehouse tools, Microsoft Azure data warehouse tools, and Google data warehouse tools. If you want to try practicing with these tools, I suggest you find one (I like Snowflake), and look up tutorials on that software specifically.

Enough about storing data—let's talk about how to update data that has already been stored.

Updating stored data

Sometimes information changes, and you must update your dataset. In such cases, there are decisions that must be made. Each decision has pros and cons based on the reason you are collecting the data. For slowly changing dimensions, two cases often come up:

- Updating a current value
- Changing the number of variables being recorded

Updating a record with an up-to-date value

In most cases, you will simply add data points to the end of a table, but sometimes there is a specific value that is calculated or recorded that you need to keep as up to date as possible. Now, you have two options:

- **Overwrite historical values**: If you just change the value in the cell, this keeps your dataset much smaller and simpler. However, because you have lost what your value was, you no longer have access to historical data. Historical data has many uses and is required for trend analysis. If you don't care about predicting future values of this number and just want to keep your dataset as simple as possible, this is the appropriate path.

- **Keep historical values**: Keeping historical values usually requires the addition of extra columns so that you can keep track of what the current value is, as well as when other values were active.

To do this, you add the following columns:

- **Active Record**
- **Active Start**
- **Active End**

These columns work in a direct manner. **Active Record** simply states whether the specified value is the most current value, and is either **Yes** or **No**. **Active Start** describes the date a value became active, and **Active End** describes the date it stopped being active. This is shown in *Figure 2.7*:

Magic Number	Active Record	Active Start	Active End
41	Yes	11/11/2011	

Figure 2.7 – Active Record

If a number is currently active, **Active Record** will be **Yes** and it will have an **Active Start** date, but no **Active End** date:

Magic Number	Active Record	Active Start	Active End
41	No	11/11/2011	12/12/2012
42	Yes	12/12/2012	

Figure 2.8 – Updated Active Record

As shown in *Figure 2.8*, when you are updating this value, you change **Active Record** to **No** and add an **Active End** date for the old value. The new value will set **Active Record** to **Yes** and receive an **Active Start** date. In this way, you can keep a record of all historical values. If you are more interested in being able to perform trend analyses and predict a future value using historical values than you are in keeping your dataset small and simple, this is the appropriate approach.

Changing the number of variables being recorded

Occasionally, you will need to change the number of variables being recorded, so you are adding or removing columns from a table or dataset. Whether you are adding or removing variables, you will have to decide whether or not to delete all historical data. It may seem odd, but there is a specific reason: null values, or spaces where there are no values where there should be. It doesn't matter whether you are adding columns or removing them—either way, you will create null values.

In *Figure 2.9*, we are adding columns that track the color of beans, which was not tracked until the third data point. This means that all data points before you added the columns will have null values:

Total Number of Beans	Red Beans	Blue Beans	Yellow Beans
10	X	X	X
12	X	X	X
11	3	5	3
9	2	6	1
10	4	3	3

Figure 2.9 – Adding variables

In *Figure 2.10*, we are no longer tracking the color of the beans, so we are removing them as a variable, and normally we would just delete the columns. However, since the historical data—the first two data points—still has these columns, you will have null values for everything after you stop tracking these variables:

Total Number of Beans	Red Beans	Blue Beans	Yellow Beans
10	3	4	3
12	2	6	4
11	X	X	X
9	X	X	X
10	X	X	X

Figure 2.10 – Removing variables

The only way to completely avoid these null values is to dump the historical data for those columns. That said, sometimes you can't or just don't want to get rid of so much data, so you will have to address the null values by other means. *Figure 2.11* is an example of a table that has had the null values removed:

Total Number of Beans	Red Beans	Blue Beans	Yellow Beans
10	3	4	3
12	2	6	4
11	3	5	3

Figure 2.11 – Deleting historical values

Okay—now you know about updated stored records. Next, we will jump into different data types and file types so that you can know what to expect from each.

Going through data types and file types

Data comes in countless formats, each requiring different treatment and capable of different things. While each programming language has its own data types, these will not be tested because the exam is vendor-neutral and does not require knowledge of any specific programming language. However, there are some generic data types that everyone working with data should know that are covered in the exam.

Data types

When discussing data types, we are talking about the format of specific variables. While there are some commonalities between programming languages, these data types may have different names or be subdivided into different groups. However, all data processing programs should have the following data types:

- Date
- Numeric
- Alphanumeric
- Currency

Date is a data type that records a point in time by year, month, and day. This data type can also include hours, minutes, and seconds. There are many different ways to format a date variable, but the **International Organization for Standardization (ISO)** recommends ordering your dates from the biggest unit of time to the smallest, like so:

- `YYYY-MM-DD`
- `YYYY-MM-DD HH:MI:SS`

That said, it is more important to be consistent within a dataset than to have any one particular format. When merging two datasets from different sources, check to make sure they are both using the same format for any date variables.

Numeric data is made of numbers. Different programs break this up into multiple different subtypes, but for the exam, all you need to know is that a value that is a number, no matter whether it is a decimal or a whole number, is considered numeric data.

Alphanumeric data includes numbers and letters. Just as with numeric data, these go by many different names, based on the program used, but include any value that has letters in it. The only exception is if the value is a number formatted in scientific notation.

Currency data includes monetary values. This one is pretty simple. Just remember that if the numbers show a dollar sign, it is counting money and is probably formatted as currency.

> **Note**
> The exam does not cover Boolean values or values that can only be TRUE or FALSE. Not every program recognizes Booleans as their own data type, and even if the program does, these values are often translated into a different data type for use. For example, a Boolean might be recoded as 1 and 0 instead of TRUE and FALSE, so it will be recognized and processed by a **machine learning (ML)** algorithm, most of which require specific data types.

Variable types

When discussing variable types in the context of this book, we are really looking at different kinds of statistical variables. What that means is if the exam indicates a specific column in a spreadsheet, you will have to be able to tell whether it is discrete, continuous, categorical, independent, or dependent. These are the types of things you will need to know when working as a data analyst to figure out whether you can run an analysis or not because every analysis has specific data requirements.

Discrete versus continuous

Discrete and continuous are two different kinds of numeric or currency data. Discrete variables are counts and usually describe whole numbers or integers. There are limited possibilities that a discrete variable can have.

Discrete examples:

- 2
- 37
- $1.23

> **Important note**
> Sometimes a number can be a decimal and still be discrete. For example, when counting currency, $1.23 can still be considered discrete, because the values after the period represent cents that can be counted individually. However, $1.235 would no longer be considered discrete because there is no way to count half of a coin.

Continuous variables are not limited to whole numbers and can represent an infinite number of values between two points. Often, these values are measured or calculated and are represented as decimals.

Continuous examples:

- 3.47
- 7.00
- $1.235

Categorical

Categorical variables, sometimes called dimensions, represent classifications or groups. Often, these are formatted as alphanumeric. There are three main types of categorical variables:

- Binary
- Nominal
- Ordinal

Binary variables are categorical variables that only have two possible states, such as TRUE and FALSE, 1 and 0, Success and Failure, or Yes and No. All Booleans are binary variables.

Nominal variables are categorical variables that contain more than two groups and have no intrinsic order. The majority of categorical variables fall into this classification. Nominal variables can include things such as color, breed, city, product, or name.

Ordinal variables are categorical variables that have an intrinsic order. These are most often represented as scales. Ordinal variables can include things such as Small, Medium, and Large or Low Priority, Medium Priority, and High Priority.

Independent versus dependent

Independence is one of the most important distinctions in statistics and will be featured heavily in the Data Analysis domain of the exam. You can consider this the purpose of a variable in a study.

Independent variables are the variables in a study that you are manipulating directly. These variables are independent because they are not influenced by anything besides you. Independent variables cause changes in other variables (or don't).

Dependent variables are the variables in a study that you are measuring. You do not manipulate these variables at all. If these variables change, it is because of the independent variables, so the values of these variables are dependent upon the values of the independent variables.

Let's look at an example. You run a simple study where you want to find out whetherbeing able to see impacts the accuracy of dart throwing. You gather twenty people, blindfold ten of them, have each of them throw three darts, then measure the distance those darts landed from the center of the target. In this example, the variable of sight, or whether the person was blindfolded or not, is your independent variable. You are directly manipulating this variable by choosing who to blindfold. Your dependent variable is what you are measuring. In this case, you are measuring the distance of the darts from the center of the target. After all of this is done, you will run an analysis to see whether changing your independent variable had an impact on your dependent variable.

File types

Often, data is saved on a computer as a file. Different types of data are saved as different types of files. A data analyst may be expected to deal with any number of file types, so the exam tests to see whether you can identify what kind of information can be found in the most common file types. The exam includes the following:

- Text (such as TXT)
- Image (such as JPEG)
- Audio (such as MP3)
- Video (such as MP4)
- Flat (such as CSV)
- Website (such as HTML)

> **Important note**
>
> While the exam is vendor-neutral and tries to avoid file types that require you to know a specific software, some of the file types are associated with a particular **operating system (OS)** because they are common enough that you are likely to encounter them if you are dealing with the associated data type. Also, these file types, while more common on some OSes than others, can be played on any OS. For example, a **WMA** file type is short for **Windows Media Audio**, so it is inherently associated with the Windows OS. That said, you do not need to be an expert in this OS to remember that a WMA file is an audio file.

Text

Files that only contain text may be common, depending on the specific data analytics position. Many word processing programs have their own file type. However, since the exam is vendor-neutral, we need only discuss ones that are not inherently associated with any particular program. For text files, that leaves a plain text file:

* **Text (TXT)**

Image

Images have several different file types that store the image in different ways. The most common include the following:

* **Joint Photographic Experts Group (JPG/JPEG)**
* **Portable Network Graphics (PNG)**
* **Graphics Interchange Format (GIF)**
* **Bitmap (BMP)**
* **Raw Unprocessed Image (RAW)**

Audio

Audio files include the following popular formats:

* **MPEG-1 Audio Layer III (MP3)**
* **Waveform Audio (WAV)**
* **Windows Media Audio (WMA)**
* **Advanced Audio Coding (AAC)**
* **Apple Lossless Audio Codec (ALAC)**

Video

Video files include the following popular formats:

- **MPEG-4 Part 14 (MP4)**
- **Windows Media Video (WMV)**
- **QuickTime Video Format (MOV)**
- **Flash Video (FLV)**
- **Audio Video Interleave (AVI)**

Flat

Flat files contain a simple two-dimensional dataset or spreadsheet. Again, every spreadsheet software has its own unique file type, but for the purpose of this exam, there are only two generic file types you need to know:

- **Tab-Separated Values (TSV)**
- **Comma-Separated Values (CSV)**

The difference between these two is how they separate values, or which delimiter they use.

TSV values are separated by tabs. Here's an example:

```
Column1      Column2      Column3
```

CSV values are separated by commas. Here's an example:

```
Column1,Column2,Column3
```

Website

When discussing website file types, we are talking about file types that can be used by a website to store or convey information to be used by a data analyst and not specifically file types used to create or manage websites. It should also be noted that these file types do represent specific languages, but you do not need to know these languages to extract information from them. For example, you do not need to understand how to structure a website with HTML to extract useful information from an HTML file; there are parsers that can do this for you. The types of website files recognized by the exam are as follows:

- **Hypertext Markup Language (HTML)**
- **Extensible Markup Language (XML)**
- **JavaScript Object Notation (JSON)**

HTML is a common file type that focuses on website structure, and occasionally passing information. Information is stored between tags. The tags create elements that all have specific pre-determined meanings and act in specific ways when used. Here's an example:

```
<div>
     <h1>
           Store Data Here
     </h1>
<p>
           Or Here
     </p>
</div>
```

XML is similar to HTML, but the tags have no pre-determined meanings and don't in a specific way. You can use whichever tags have meaning to you. For this reason, it can be difficult to parse information from an XML file that came from a new source. Here's an example of XML:

```
<Dataset>
     <Data>
           Store Data Here
     </Data>
<AlsoData>
           Or Here
     </AlsoData>


</Dataset>
```

JSON files are not used to structure websites, unlike the other two. JSON specializes in storing and passing information. A JSON file contains a list of data objects and gives values to those objects using key-value pairs. Here's an example:

```
"Dataset" : [
    {
          "Data" : "Store Data Here"
    },
    {
          "Data" : "Or Here"
    }


]
```

In the end, you do not need to know how to use any of these languages. Make sure you understand that while all three can pass information, JSON is the one that specializes in it. Also, know that JSON is the only one that does not contribute to website structure and is based on key-value pairs.

Summary

We covered a lot of information in this chapter. First, we covered structured and unstructured databases, and what types of data can be expected in each. Also, we talked about relational and non-relational databases, and how they relate to structured and unstructured databases. Next, we covered database schemas such as star and snowflake schemas. Then, we covered data warehouses, data marts, and data lakes. Briefly, we touched on how to update stored data. Finally, we wrapped things up with different data types and file types. This is everything you need to know about the storage of data.

In the next chapter, we will go over how this data is collected in the first place!

Practice questions and their answers

Let's try to practice the material in this chapter with a few example questions.

Questions

1. A smart thermometer collects information about the temperature outside every 30 minutes, creates a log, and sends the data to a local database. What can you tell about the database?

    A. It is structured

    B. It is unstructured

    C. It is relational

    D. There is not enough information

2. Client-facing agents at banks are not technical experts, but require the ability to query client information. Which data schema is most appropriate for their database?

    A. Star schema

    B. Snowflake schema

    C. Galaxy schema

    D. Avalanche schema

3. You are working as a data scientist for a video-streaming website. You require access to raw, unprocessed data and video files. What is the most appropriate database style to use?

   A. Data warehouse

   B. Data mart

   C. Data lake

   D. Data mine

4. An e-commerce website has been collecting information on purchases for over a year. Now, they want more detailed geographical information to find out where their products are selling, so they start collecting information on the IP address of the person who made the purchase. This adds a new column to the dataset. What concerns might they have going forward?

   A. Historic values for the IP Address column will be null

   B. New values for the IP Address column will be null

   C. They will not know which record is active

   D. None of these

5. You are given a file with the ".png" extension at the end to process. What type of data will you find inside?

   A. Video

   B. Audio

   C. Text

   D. Image

Answers

Now, we will briefly go over the answers to the questions. If you got one wrong, make sure to review the topic in this chapter before continuing:

1. The answer is B. *It is unstructured*

   The question tells you that the data is automatically generated and logged by a machine. This makes it machine data, and machine data is one of the two types of data that are inherently unstructured.

2. The answer is A. *Star schema*

   A star schema is simple and focuses more on being user-friendly, making it ideal for data marts and use by less technical employees.

3. The answer is *C. Data lake*

   A data lake is the only database style discussed that focuses on raw data. It is also the only one that can easily store structured and unstructured data or is meant for use by a data scientist.

4. The answer is *A. Historic values for the IP Address column will be null*

   Because you are adding a new column, you will not have any values in that column from before you added it, making all historic values null by default.

5. The answer is *D. Image*

   PNG stands for Portable Network Graphics and PNG files contain images.

3
Collecting Data

In this chapter, we will discuss sources of data. Sometimes data is not already nicely and neatly stored for you, and you will have to reach out to collect your own. Here, we will discuss the most common sources of information and a couple of systems that will pull or store data automatically. You will also learn how to make your queries more efficient if you are accessing a database. These resources and techniques are crucial, not only to passing the CompTIA Data+ certification exam but also to the everyday life of a data analyst.

In this chapter, we're going to cover the following main topics:

- Utilizing public sources of data

- Collecting your own data

- Differentiating **Extract, Transform, Load** (ETL) and **Extract, Load, Transform** (ELT)

- Understanding **online transactional processing** (OLTP) and **online analytical processing** (OLAP)

- Optimizing query structure

Utilizing public sources of data

There are all kinds of datasets available for free. With these, collecting data is as simple as asking for it and downloading it. There are all sorts of information out there on any number of topics, free for anyone to use. Some of these free sources are even already clean and organized, just waiting for an analysis to be run.

Public databases

Public databases are databases that legally must be accessible. These fall into one of two categories: **government** or **industry**. The majority of public databases are run by government entities. These can be found around the world, sometimes at local and national levels. What information is specifically shared depends on the government entity, but can include information about the following:

- Population

- Agriculture

- Utilities

- Public health concerns

The downside to government-run databases is that, while they are free, they are usually full of missing data and mistakes. This means that you will have to spend a lot of time cleaning the data.

The public data you have access to depends entirely on where you are in the world. Each country may make data accessible to its own citizens but may try to limit access to citizens of other countries. While there may be *creative* ways around these limitations, getting too "creative" is not considered the best practice. Alternatively, some international organizations allow access to their data in much the same way as a government agency. The **World Health Organization** (**WHO**) makes all kinds of data accessible. You can visit `who.int/data/collections` and give it a try.

Industry-run public databases are about specific industries that are legally mandated to release specific data. These are generally cleaner than government databases, but they tend not to share more than the legal minimum, and they do not make this data easy to find.

Open sources

Open source datasets are ones that have been posted by individuals or companies for free. This data is freely given and covers a very wide range of topics. There are several of these sources in existence, but it is generally accepted that `www.kaggle.com` is the most useful.

Based on a community, there are datasets posted from individuals, companies, universities, and organizations. Over 50,000 datasets have been posted, and you can filter your search by file size, file type, licenses, or topics.

> **Important note**
> Let me be explicit here—being able to filter by license is important because not all of the datasets can be used commercially. If you are just creating a portfolio project to publish on Kaggle, it doesn't matter, but if you want to try to use the data for your company, you will have to check the licenses carefully.

Kaggle also offers courses and competitions; it is a great place to get started in the data analytics community. To be clear, Kaggle is not the only resource like this. A lot of people like Dataworld, which offers free personal accounts, but to use the information commercially, you need to purchase an enterprise license. You can learn more about this at `data.world/pricing`. Because the data is coming from so many sources, it is not uniformly clean and ready for use, but it tends to be cleaner than data in government-run public databases.

Application programming interfaces and web services

Some companies will allow limited access to their databases remotely using an **application programming interface (API)**. This limited access is different from the public sources that allow you to just download an entire dataset outright. There are different types of APIs, and they work in different ways, but you can generally think of them as a piece of code that allows two unrelated computer systems to exchange information. For example, you go to a restaurant and sit at a table. You tell the waiter what you want, and he goes to the kitchen to tell the cook your order before bringing the food back to you. In this case, the waiter is the API, the cook is the database, and the food is the data you want. The API is simply a middleman that allows you to communicate with a database.

Web services are a specific kind of API that uses a hosted network and require both computers to be in the same hosted environment to work. That said, with small protocol differences, they still work like other APIs. For the purpose of the exam, you need to know that all web services are APIs, but not all APIs are web services.

All APIs pass information back and forth in one of two ways:

- Synchronous

- Asynchronous

When requesting information through an API by synchronous means, the code will make the request and then wait for a response. An asynchronous API request means that the code continues to run while it waits for the API to return with the data. The advantages are straightforward; with asynchronous code, you can be faster and more efficient because you do not have to stop everything and wait. However, the downside is that if you don't wait, your code may try to use the data before the API responds with it, which will cause an error in your code.

> **Important note**
>
> Because the exam is vendor-neutral, you do not have to know any specific programming language. This means that you will not be asked to create an API or to make a request to an API. Instead, you will simply have to have a firm understanding of what they are, why they are used, and the difference between synchronous and asynchronous.
>
> That said, APIs are very useful resources. Most data analysts do not use them in their everyday job, but they are commonly used to acquire data for portfolio projects. While you may never be asked to build an API, it does not hurt to learn how to request data from one, just so you are comfortable with the process.

Thousands of APIs available connect to databases with all sorts of information, including entire databases of bad jokes. That said, many major tech companies have APIs available, including Facebook, Amazon, Apple, Netflix, and Google. There is a lot of good information out there available using APIs. The downside of the APIs is that they are not uniform. Every API is set up slightly differently; you will have to read the documentation and, in most cases, acquire authorization from the company in order to request information. Several APIs also have limits on the size or number of requests you can make to ensure that one person isn't bogging down the entire system for everyone.

If you would like to learn more, try one out. A popular one for people just getting started is the Twitter API, which you can check out at `developer.twitter.com/en/docs/twitter-api`. This API is popular because it is well documented, there are lots of guides out there on how to use it, and it gives access to a lot of data. You can even go to the website and try it by clicking the **Try a live request** button, without writing a bunch of code, though you do need an account. The specifics on how to connect to it, how to get authorization, how much data you can pull for each level of authorization, and everything you may want to know is given in the documentation.

Public data sources are a great resource, and it never hurts to look through them first, but sometimes you just need to collect your own data.

Collecting your own data

You will not always be able to find the exact dataset you require nicely collected and cleaned for you with a little bow on top. Often, if you need specific data, you will need to go out and get it yourself. This can be easier said than done. However, there are several different ways for you to collect your own data. We will go over a few of the most common approaches you will need to know.

Web scraping

As the name implies, **web scraping** is the process of collecting data from the web. While it uses the web, this approach is different from public databases, open sources, or APIs because there is no database. Web scraping is the process of collecting information directly from a web page instead of a database. This can include collecting price information on a product from several different sites, collecting posted stock information, or filtering through social media posts with specific tags. Anything that is posted on a website can be accessed and saved in your own dataset. To be clear, you are not visiting a site and writing down the information on a notepad. The code will automatically go to a specified location, collect specific information, and bring it back.

Web scraping is a useful tool and has many applications. A lot of information is posted on the web in one form or another, and web scraping allows you to collect it into something useful. The problem that a lot of people face is that to do it well requires you to understand how web pages are structured well enough to be able to target the specific information you want. More often than not, this means that the data analyst has to at least understand the basic concepts of HTML in addition to whichever scripting programing language they are using to run the web scraping code. For more information on website structure, look up guides on the **Document Object Model (DOM)**.

Surveying

Surveys are one of the most common ways to collect information. At its heart, a **survey** is just a set of questions that you give to a sample of individuals. A sample is a small subset of a larger population.

> **Important note**
> A population represents every single individual in a group. If you want information on people who own pets, the population is everyone on the planet who has a pet. Getting information on everyone is not possible. However, you take information from a sample and generalize what you learn from that sample to the population as a whole. This is a fundamental principle of inferential statistics. If you want to learn more, look up a guide on sample techniques.

These can be done electronically, with paper forms, or by someone simply asking questions and writing down the answers. While you can get just about any kind of data from a survey if you do it properly, they are often used to collect information on demographics and customer satisfaction.

Types of survey answers

There are many ways to ask the same question, each with its own set of pros and cons. For a moment, we will focus on the types of survey answers. The main types are as follows:

- Text-based
- Single-choice
- Multiple-choice
- Drop-down
- Likert

Text-based answers are exactly what they sound like. You provide a blank field and let the person taking the survey write in whatever they want. From a data analytics point of view, these are a pain. There is no way to control how the person answers, so you never know what you will get. Best-case scenario, you can try to use **natural language processing (NLP)** to automatically give values to it. More likely, you will have to go through and look at every value individually, which can take time if you have a big sample. That said, there are certain types of data you can only collect through open-ended questions such as this:

1. What is your favorite kind of peanut butter? ⚲ 0

[   ]

Figure 3.1 – Text-based survey question

In *Figure 3.1*, we see an example of a text-based survey question. It is simply a question, with an empty box under it where someone can write a response.

Single-choice answers are where a list of answers is provided and the person taking the survey can only pick one. These are clean and the results are easy to analyze. An example is shown here:

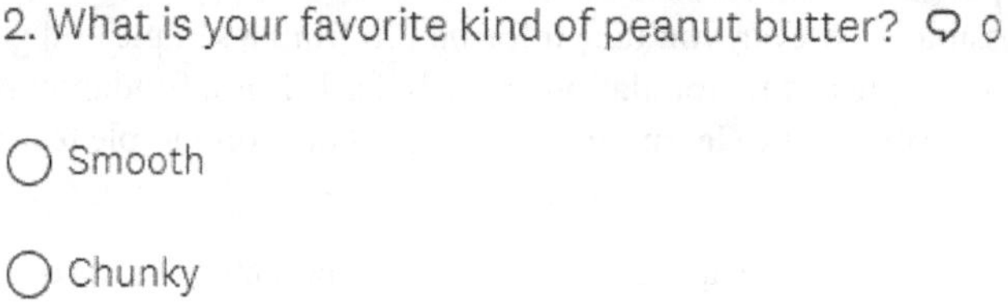

2. What is your favorite kind of peanut butter? ⚲ 0

○ Smooth

○ Chunky

Figure 3.2 – Single-choice survey question

In *Figure 3.2*, we see the same question as *Figure 3.1* but formatted as a single-choice answer. The format is denoted by having circles next to the possible answers so that only one can be selected.

Multiple-choice answers are where you provide a list of possible answers and tell the person to select all that apply. They can click as many or as few as they like. These can take a little more time to analyze than a single-choice answer, but they are not complicated. Here's an example:

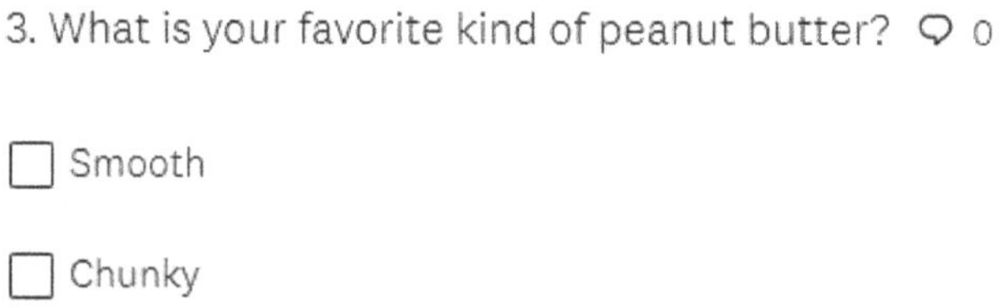

3. What is your favorite kind of peanut butter? ⚲ 0

☐ Smooth

☐ Chunky

Figure 3.3 – Multiple-choice survey question

In *Figure 3.3*, we see the question written as a multiple choice. This is denoted by the squares next to the answers, suggesting that more than one answer can be selected. Sometimes, this is even stated after the question with a "*Select All that Apply*" statement.

Drop-down menus are where the person taking the survey selects a value from a drop-down list. From an analyst's point of view, the results of these are almost identical to a single-choice answer. Either way, every person who takes the survey will select one of a few possible outcomes. Here's an example:

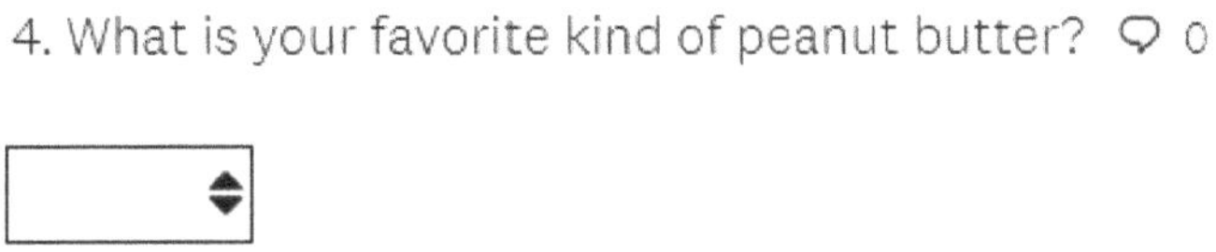

4. What is your favorite kind of peanut butter? ♡ 0

Figure 3.4 – Drop-down survey question

In *Figure 3.4*, we see the question formatted as a dropdown. This is denoted by the arrow symbols in the box. Here, the person taking the survey would click on the box and then select an answer from a drop-down menu.

Likert scales are popular, specifically when gauging the popularity or effectiveness of a specific part of a product. These are the questions that make a statement, and the person has to answer "Strongly agree", "Agree", "Neither agree nor disagree", "Disagree", or "Strongly Disagree". Again, the results are clean and easy to analyze. In addition, Likert scales are great for turning qualitative questions into quantitative results.

> **Important note**
>
> Qualitative data refers to specific qualities that describe something. Color is a quality, so saying *the car is red* is qualitative data. Quantitative refers to things that can be quantified or things that can be counted, measured, or calculated. These are your numbers, so saying *the car has four tires* is quantitative data. There is a time and a place for both kinds of data, and ways to transform qualitative data into quantitative data and quantitative data into qualitative data.

Now, let's look at a quick example:

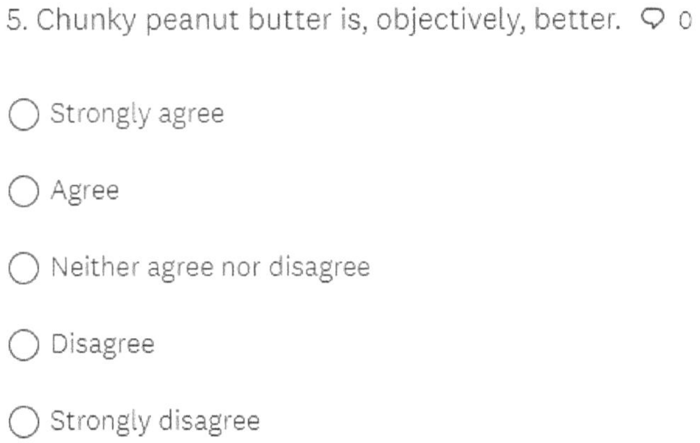

5. Chunky peanut butter is, objectively, better. ♡ 0

○ Strongly agree

○ Agree

○ Neither agree nor disagree

○ Disagree

○ Strongly disagree

Figure 3.5 – Likert survey question

In *Figure 3.5*, we see the same question formatted as a Likert scale. Note that this is set up as a single-choice question, but the difference is that the question is written as a statement and the possible answers are written as a scale. There are multiple ways to set up a Likert scale, but this is one of the most common.

Each of these types of survey answers is appropriate for different kinds of questions. It all depends on what you want to know.

Survey bias

Bias is the bane of analysts who don't pay attention to actively avoiding it. To put it simply, **bias** is the difference between the expected value and the actual value. Bias means that what you learn about your sample cannot be generalized to your population. The more bias you have, the further your results will be from the truth.

Okay—you understand that bias is bad and invalidates your results, so how do you avoid it? That is tricky because there are hundreds of different kinds of biases, and there is no surefire way to avoid all of them. Randomizing who receives the survey as much as possible is a good start, but there are a few types of bias that are common to surveys that are easy to avoid.

> **Important note**
>
> For more information on specific types of bias, check out www.thedecisionlab.com/biases. This is a great source that covers the most common kinds of bias and how to avoid them.

Order bias is common in surveys, and it happens when the order of the questions or the order of the answers impacts what people choose. This is easily avoided by randomizing the order of the questions and answers. Most survey sites will include this randomization as an option or automatically mix things up for you.

Also, try to avoid leading questions or answers. For example, you shouldn't ask, "Which do you prefer: The incredible product A or the lackluster product B?" Leading questions, whether intentional or accidental, add bias. Instead, try to avoid using adjectives at all.

Recall bias happens when you ask someone to remember details of something that happened in the past that they may or may not remember. This is especially problematic if you are requesting a text answer to the question. Not able to clearly remember what happened, the person is likely to just guess, which, as you can imagine, does not always yield accurate results. Try to avoid asking questions about the distant past, especially with text answers. If you must ask about the past, add "I don't know" as a possible answer so that they don't have to make something up.

If you are interested in surveys, I highly suggest going to a service such as SurveyMonkey, at surveymonkey.com, signing up for a free account, and creating one. There are even guides and suggestions on the survey-making process.

Pass the survey around to friends and family and just see what kind of results you get. It will help get you familiar with the process and get you some practice writing useful questions. SurveyMonkey also has templates, suggestions, and people you can talk to about your survey.

Overall, making a survey is not hard, but making a survey that will deliver accurate and useful results is not easy. There are entire companies that specialize in creating and delivering good surveys if you aren't comfortable doing it yourself. If you want to learn more, look up guides on survey design.

Observing

Observation can be very simple or incredibly complicated. At its heart, observation is simply witnessing something and recording it. The process of devoting time and effort to gather and record specific observations is called a **study**.

> **Important note**
> For more information on specific types of studies, look for guides on study design. If you are feeling bold, or particularly academic, look for guides on research design.

There are many kinds of studies. Some studies are passive, where you simply watch and see what happens. Other studies are active, where you change something in the environment and see how other elements react. The specifics of the study depend entirely on the type of data you want to get out of it.

More importantly, there are different ways to observe things. The classic approach is physical observation, which is the process of going to a location and recording what you see. For certain types of data, this is required. If a department store wanted to know how many people who go to the mall wear hats, you would need to physically go to the mall and record the number of hats you saw. This process takes time but is the best approach for studies that require decision-making. Does a visor count as a hat? Is that lady wearing a bonnet?

A common example is an A-B study for a website. The basic design is roughed out in *Figure 3.6*:

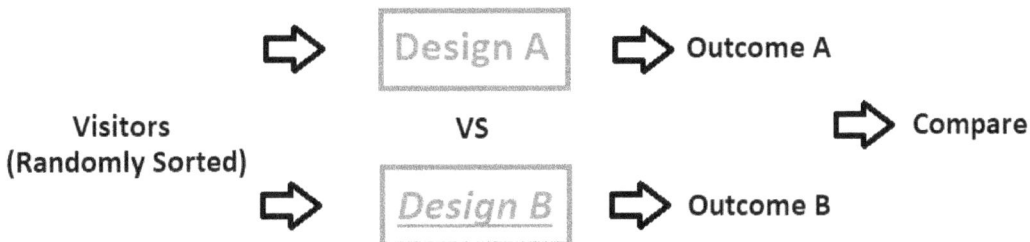

Figure 3.6 – A-B study for a website

A company is trying to decide which website design leads to more sales: design A or design B. Every customer who visits the website is randomly routed to either design A or design B. The website automatically observes and takes notes of things such as how long a person stays on a page, how many links they click, or whether or not they buy something. Once enough data is collected for each design, you can run a statistical analysis to see which design performed better with a specific goal in mind.

More and more, automated observations are becoming popular, and software will generate metrics for you. A website will count how many times someone looks at a specific page, or how many times a link is clicked. The department store can set up a program to count the number of hats sold. These are also considered observations. While you may not physically see it happen, it is a record of something that has occurred that was observed by some mechanism that was looking for it. Needless to say, automated observations are convenient but limited in the types of information that they can collect.

Differentiating ETL and ELT

You don't always have to collect your data manually. Some programs will automatically pull data from a source, prepare it for use, and move it to a new location, usually your local environment. The code to automate this process is called a **data pipeline**. There are many kinds of data pipelines, and each will need to be tuned to which data you are pulling and what you need to do with it. While more complicated pipelines can automate entire modeling and reporting processes, the exam focuses on two types, and both types have the same three steps:

1. Extraction
2. Transformation
3. Loading

Extraction is the step of pulling the data from the original source. The source can be a database you own, an outside database, or even an automated web scraping system—it doesn't matter. Extraction is picking up the information from wherever it was originally stored. This is similar to the process you would do to manually collect it yourself, but it is automated.

Transformation can take on many forms, but the idea is that this step prepares the data for use. It cleans the data and organizes it nicely and neatly so that it is ready to be plugged into an algorithm as soon as it arrives. This step can include dealing with missing values, adjusting capitalizations, normalizing data, combining columns, recoding, removing data, or any number of other operations. The idea is that this automates the step of data manipulation to save you time and effort.

Loading is the process of placing the data into a new environment. More often than not, this is the environment where you will actually be working with the data. Often, this means you are bringing it to your local machine or programming interface, but there are times when it is more appropriate to move the data to a **virtual machine** (**VM**) or a cloud environment. Sometimes, the data is loaded directly into a data warehouse.

For the purpose of this exam, you don't need to be able to create or run a pipeline. Not only would this require knowledge of a programming language, but also, the creation of a pipeline is not generally expected of someone just starting their data career. However, you will be expected to know what a pipeline is and the difference between two specific kinds of pipelines. The main difference between these pipelines is the order in which they perform the three listed steps and the consequences of changing the order.

ETL

ETL is a pipeline that extracts the data, then transforms it, then loads it to the destination. While this may seem obvious, it means that the data is transformed after it is extracted and before it is fully loaded; the data is transformed in transit or while it is being loaded to the new location. In *Figure 3.7*, we can see how this plays out:

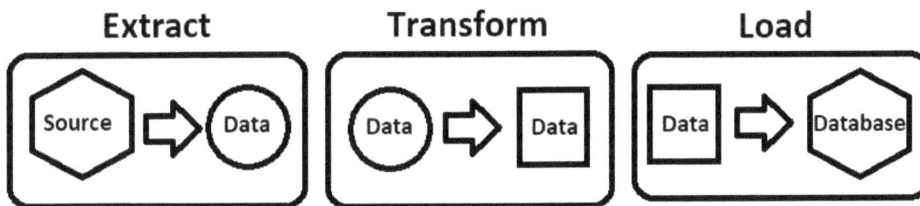

Figure 3.7 – ETL

This is quick, simple, and efficient, so long as your transformations are quick and simple. When you start to get into larger or more complicated transformations, the entire process slows down very quickly.

ELT

ELT is a pipeline that extracts the data, loads it to the destination, then transforms it. This means that the processing power to transform the data comes from the destination environment. We can see the flow of this process in *Figure 3.8*:

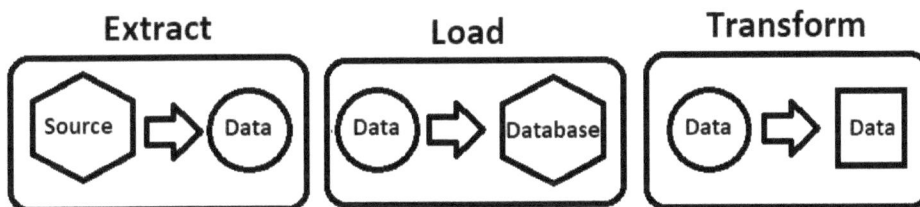

Figure 3.8 – ELT

Several companies have taken a liking to this approach because they can load the data to a VM in the cloud and then pay to rent the processing power they need to transform the data quickly. This negates the requirement to own high-powered hardware to work with large or complex data.

Generally speaking, ELT loaded into a rented environment is faster than ETL when working with complicated transformations, but it is more expensive. For individuals or small companies with simple transformations, ETL is still the preferred approach.

Delta load

It is important to mention that data can be loaded in different ways. These loading methods exist with or without a pipeline, but they do have an impact on the efficiency of an ETL or ELT pipeline. The most important approaches are as follows:

- Full load
- Delta load

To keep it simple, a **full load** means that every time the pipeline is run, the entire dataset is extracted, transformed, and loaded. Occasionally this is performed as a safety measure in the form of redundancy, but it is not efficient.

In scientific terminology, delta (Δ) represents change, so a **delta load** only loads things that have changed since the last load. As you can imagine, it is much faster and more efficient to only update things that have changed instead of trying to load the entire dataset every time.

Understanding OLTP and OLAP

OLTP and OLAP are relatively new, but important, fields. They involve the financial transactions performed over a network, such as shopping or banking transactions. To be clear, these are a kind of automated observation and often include a pipeline such as ETL or ELT, but are considered their own categories because financial data is protected. Some guidelines must be followed for OLTP and OLAP that are not required for other types of data. Specific guidelines on protected data will be covered later in this book.

OLTP

OLTP is the act of automatically storing and processing data from online transactions. To be clear, that means that every time you purchase something online, the record is automatically collected, stored, and processed through OLTP. More often than not, this process uses a snowflake schema, as discussed in *Chapter 2, Data Structures, Types, and Formats*. There are specific safeguards involved with OLTP, such as making sure that if some part of the transaction fails, the entire process stops. Another safeguard is that a specific piece of data cannot be changed by two users at the same time, which means everything happens one after the other. Both features are there to ensure data integrity.

OLAP

OLAP is the next step. After the data has been collected and stored by OLTP, OLAP takes the information, often through a pipeline such as ETL or ELT, and moves it to a new database, or data warehouse. The new database is often in a star schema, as discussed in *Chapter 2*, *Data Structures, Types, and Formats*. This process includes simple analyses and aggregations so that the data is ready to help people make informed business decisions with minimal work.

OLTP and OLAP sound very similar. To keep them separate in your mind, just remember that A is for analytical, so OLAP is the analysis step. By process of elimination, OLTP must be the collection step.

Optimizing query structure

A query is simply a request for information, so in the field of data analytics, a query is a request for data. It is how we call data from a database to our local environment. Several different programs allow you to query a database, and some are more popular than others, but often the biggest difference in the performance of queries is decided by how optimized they are. If you are making a simple query to a small database, efficiency and performance may not mean very much, but as the amount of data you are pulling gets bigger and your queries get more and more complicated, performance becomes a bigger issue. When you get to the point where a query takes hours or even days to run, you might want to consider optimizing it.

Filtering and subsets

First and foremost, it should be obvious that the less information you pull, the faster the process will be. **Filtering** is the process of being selective about which data you are querying. There are many ways to filter data, and it depends on what program you are using. However, at its core, a filter uses conditional logic. You can say that you want data from a particular table if it meets certain conditions. Often, you are filtering for specific values or a range of values. Again, the more specific a query is, the less unnecessary data you are pulling and the less processing power you are wasting.

Subsets are exactly what they sound like: a smaller subsection of your dataset. Filtering effectively creates a subset. The important thing to remember is *when* you use a filter to create a subset. It is vastly more efficient to create a subset before you join tables. Squishing tables together takes a fair amount of processing power. If you add a filter after you create a join, then you have to join a much larger table and it takes more time. Filtering data before joining tables means that you are only joining a subset of the whole table, which is much more efficient.

Have a look at the following screenshot:

Employee ID	LastName	FirstName	Department	YearsWithCompany
83784	Benhill	Floyd	Sales	12
64986	Chane	Jill	IT	1
93671	Hanson	Richard	HR	15
37816	Smith	Trudy	Sales	21

Employee ID	LastName	FirstName	Department	YearsWithCompany
83784	Benhill	Floyd	Sales	12
37816	Smith	Trudy	Sales	21

Figure 3.9 – Filtering

In *Figure 3.9*, we see a simple dataset that has had a filter applied. In this case, the filter is simply selecting all employees with the Sales values in the Department column, but filters can be applied to any of the columns and can include ranges as well as specific values. As you can see, the table including only the required data, WHERE Department = Sales, is much smaller than the original table. Using filters and subsets to only use the data you need is a great way to optimize a query.

Indexing and sorting

Indexing is assigning every entry in a table a unique ascending number, so you can count down the rows: 1, 2, 3, and so on. Why is this useful? Because the index acts as a placeholder. It means that you can temporarily sort the table and then put things back where they were afterward. You can sort without indexing first, but it is much less efficient.

Sorting is simply arranging the rows in a different order according to some logic. The idea is that, when working with a filter, sorting can save time. For example, if you only want to look at rows where the value of column X is greater than 5 and less than 10, you sort by column X, then filter it. This means that all the rows you want are right next to each other in the table. The values are easy to find and grab.

That said, there are debates about how useful sorting is. The problem is that when you are working with a large dataset, the act of sorting itself uses a lot of processing power, so you have the potential to make things less efficient instead of more. However, for the exam, you simply need to understand what sorting and indexing are and that they can play a role in query optimization.

Have a look at the following screenshot:

Index	Employee ID	LastName	FirstName	Department	YearsWithCompany
1	83784	Benhill	Floyd	Sales	12
2	64986	Chane	Jill	IT	1
3	93671	Hanson	Richard	HR	15
4	37816	Smith	Trudy	Sales	21

Index	Employee ID	LastName	FirstName	Department	YearsWithCompany
4	37816	Smith	Trudy	Sales	21
3	93671	Hanson	Richard	HR	15
1	83784	Benhill	Floyd	Sales	12
2	64986	Chane	Jill	IT	1

Figure 3.10 – Indexing and sorting

In *Figure 3.10*, we see that an index has been added to the dataset, then the dataset was sorted by `YearsWithCompany` descending. Now, the names are in order of who has been with the company the longest, but it would be easy to return them to the original order using the index. If you were looking for people who have more than 10 years of experience, then ordering the list using `ORDER BY YearsWithCompany DESC` will put those with the most experience on top together. This makes finding and filtering them much faster.

Parameterization

Parameterization is a prewritten query that allows the user to enter specific parameters to tell the query which data to target. This means that you don't need to create a query from scratch every time you want to pull data from the database. It also allows you to use a pre-optimized query without having to think about it.

The biggest reason for parameterization is not query optimization, but cyber security. By forcing people to use prewritten queries and limiting their options for what they can write, you are protecting the data from injection attacks. Malicious code can come in many forms and do many things. Limiting what people can enter is one way to protect against it.

Temporary tables and subqueries

Temporary tables are an incredibly useful tool. Effectively, you are taking the results of a query and saving them as their own table. That means that a big, complicated, time-intensive query is something you only need to run once. After that, you can run small, simple queries off this new table. If every time you ran a query you had to connect multiple tables, sort, filter, add aliases, aggregate, or any number of other things, all of that processing power would be saved if you just stored the results in a temporary table. That said, the tables are temporary, and most programs will automatically delete them after a set time. It should also be noted that they do not update when the source data updates, so they are often refreshed or recreated regularly.

An example of a temporary table might look something like this:

```
#Temporary Table
CREATE TEMPORARY TABLE EmployeeJones
SELECT * from Employees
WHERE LastName = 'Jones';
```

Again, you do not need to know a specific programming language for the exam; this is just to give you an idea. This particular example is for MySQL, and the syntax is different for different versions of SQL. Effectively, this code creates a new table called `EmployeeJones` and fills it with every employee with Jones as a last name from the `Employee` database.

Subqueries are simply a query embedded inside of another query. This means that you can pull a smaller set of data from a larger set as you are pulling it. Some professionals swear by subqueries, but, in general, they are considered less efficient than other options. For example, creating a temporary table and querying it is often considered a better approach than using a subquery for the same task. Let's look at an example:

```
#Subquery
SELECT * from Employees
WHERE DeptID in (SELECT DISTINCT DeptID from Departments WHERE
City = 'Atlanta');
```

Here is a query with conditional logic. The conditional logic references another table, but instead of trying to join the tables, it uses a subquery to stuff a whole other query into the SELECT statement. These are often used as a shortcut to avoid joining entire tables. They seem complicated at first, but become easier to understand with practice.

Once more, you do not need to understand how to perform a subquery for the exam; simply have a rough idea of what they are and why they are used.

Execution plan

Many programs that are used for querying will also have the option to show the execution plan. The **execution plan** is the program telling you how it will execute the query, sometimes accompanied by a graphical representation. An estimated execution plan will give you a rough idea of what the process should look like, while an actual execution plan is generated after a query is run and gives specifics about how long each step took. This breaks down the actual metrics for efficiency. By checking your execution plans, you will get tangible results that you can use to adjust and refine your queries. You can get a rough idea of what one looks like in *Figure 3.11*:

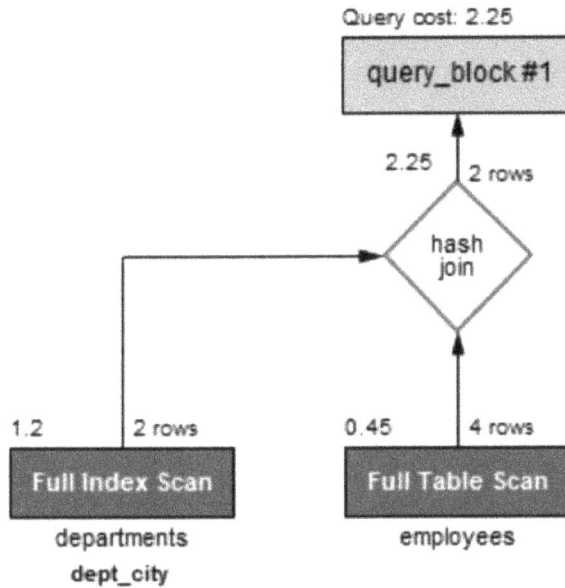

Figure 3.11 – Execution plan

The details of an execution plan will look different based on the language you are using, the tables you are using, and how you are querying them. They can be even simpler than this, or incredibly intricate and complicated. The idea is to use them to make your query as efficient as possible, while still getting the information you need.

Summary

This chapter has covered a lot of information from some pretty diverse topics. First, we talked about public sources of data including public databases, open sources, APIs, and web services, as well as the pros and cons of using each. Then, we talked about the different ways to collect your own data, including web scraping, surveying—especially the different types of survey questions and survey bias—and observations. Then, we covered the difference between ETL and ELT, as well as a full load and a delta load, and why it is important. Next, we briefly covered OLTP and OLAP and how they are used to collect and process transactional data. Finally, we wrapped up the chapter by covering ways to optimize query structures, such as filtering, subsets, indexing, sorting, parameterization, temporary tables, subqueries, and execution plans. Whew! There sure are a lot of ways to collect data. In the next chapter, we will go over what to do with it once you have it!

Practice questions and their answers

Let's try to practice the material in this chapter with a few example questions.

Questions

1. If a web service is synchronous, it means…

    A. Your system will wait for a response before continuing

    B. Your system does not have to wait for a response

    C. You are syncing a web service to your system

    D. Web services cannot be synchronous

2. When conducting a survey, it is best practice to ask about specific events that happened in the past and request a text-based answer. True or false?

    A. True

    B. False

3. ETL stands for…

    A. Extract, Transmit, Load

    B. Estimate, Time, Load

    C. Extract, Transform, Load

    D. Exact, Time, Load

4. The process of taking transactional data that has already been stored, aggregating it, and moving it to a data warehouse is called what?

    A. OLTP

    B. OLAP

    C. API

    D. Web service

5. You are provided with a long and complicated query that takes a long time to run, but you need several pieces of information from it. What is the most efficient approach?

    A. Run the query as it was given to you

    B. Add subqueries to extract specific information from the query

    C. Make sure no filters are slowing the process down

    D. Save the results of the query to a temporary table

Answers

Now, we will briefly go over the answers to the questions. If you got one wrong, make sure to review the topic in this chapter before continuing:

1. The answer is *A. Your system will wait for a response before continuing*

   Web services are a type of API. Synchronous API calls mean that your code will make a request and then wait until it gets a response before continuing to the next step.

2. The answer is *B. False*

   Requesting specific information from the past introduces recall bias and is especially bad when combined with a text-based answer. It is best practice to avoid asking about specific incidents in the past, if possible, or at least make it clear that they can give a non-answer such as "I don't know" instead of making something up.

3. The answer is *C. Extract, Transform, Load*

   ETL stands for Extract, Transform, Load because it is a data pipeline that performs these three steps in this specific order. It extracts the data, pulls it from its source, transforms it, or manipulates it into something that can be used, and finally loads it to its new location.

4. The answer is *B. OLAP*

   OLAP is the process of taking information stored by OLTP, aggregating and analyzing it, and moving it to a new database or data warehouse.

5. The answer is *D. Save the results of the query to a temporary table*

   By creating a temporary table, you can quickly and easily access anything from the results of the original query that you require without having to rerun it.

4

Cleaning and Processing Data

On rare occasions, you may receive data that is already clean, neat, and ready to use, but having an immaculate dataset just handed to you is the exception, not the rule. More often than not, while working as a data analyst, the datasets you receive will be messy, incomplete, and completely unusable without a little work. Trying to use jumbled data will only give you jumbled results. This chapter covers the most common issues you will come across and a few approaches to dealing with them.

Here, we will discuss the difference between duplicate data and redundant data, as well as what to do about it. Then, we will discuss why missing data is an issue and the different approaches you can take to deal with it. Briefly, we will cover invalid data, mismatched data, and data type validation. After that, we will discuss non-parametric data, what it is, and how to approach it. Finally, we will discuss outliers or data points that don't seem to fit in with your dataset.

In this chapter, we're going to cover the following main topics:

- Managing duplicate and redundant data
- Dealing with missing data
- Understanding invalid data, specification mismatch, and data type validation
- Understanding non-parametric data
- Finding outliers

Managing duplicate and redundant data

Often in statistics or data analytics, we are told that more data is better, but that isn't always true. If the data is duplicated or redundant, it can cause issues with skew, bias, or completely invalidate your analysis. Here, we will discuss the different ways you can have too much data, how this will impact your results, and what you can do about it.

Duplicate data

Duplicate data is when a specific data point recurs multiple times within a dataset. If we are looking at a spreadsheet, it means there are multiple rows with completely identical values. In the following table, we see a simple example of duplicate data:

Employee ID	LastName	FirstName	Department	Years With Company
83784	Benhill	Floyd	Sales	12
64986	Chane	Jill	IT	1
64986	Chane	Jill	IT	1
64986	Chane	Jill	IT	1
93671	Hanson	Richard	H R	15
37816	Smith	Trudy	Sales	21

Figure 4.1 – Duplicated data

Figure 4.1 is a bit of an exaggeration, but this does happen. In this case, Jill Chane is only one person, but she is counted three times. Okay—it's weird, but why is this a problem? Let's consider some basic descriptive statistics. As is, it looks like the IT department is the largest part of the company, taking up 50% of the employees. Without the duplicates, you would find IT is only 25% of the company, and 50% of the company is actually taken up by the sales department.

Let's look at another example in the same table: `YearsWithCompany`. This column, as you would expect, denotes how many years a specific employee has spent with the company. If you were to take the average of this column with the duplicates, you would come up with 8.5 years. However, without the duplicates, the average is 12.25 years. The percent difference between 8.5 and 12.25 is 36%; we will discuss how to calculate this later in *Chapter 7, Measures of Central Tendency and Dispersion*, but for now, realize that being wrong by 36% is a pretty big problem.

In the preceding example, the value that was duplicated was much smaller than the other values, so it skewed the data or made the mean of the dataset seem smaller than it actually was. If the duplicated value was larger than the mean, it would skew the other dataset the other way. If, however, the duplicated data is at the mean, that means you are skewing your data to the norm. This means that it makes your data look less spread out and a little nicer than it actually is. We will discuss this more when we discuss imputation later in this chapter, in the *Dealing with missing data* section. For now, understand that duplicate data can distort your data.

> **Important note**
>
> While the majority of duplicate data copies the entire row perfectly, there are times when only part of the row is copied perfectly, creating a partial duplicate. There are many different ways this can happen, but it is especially an issue in the field of marketing or customer service. In either case, if you have multiple profiles for the same person with different data, you are going to have problems.

The most common approach is simply to delete all duplicate rows. There are multiple ways to do this, depending on which software or programming language you are using. For the purpose of this exam, you do not need to know how to remove duplicate data, but you should be able to identify it, know it is a problem, and know that you should delete it. Next, we will discuss redundant data.

Redundant data

Redundant data sounds similar to duplicate data, but it is a very different problem with very different issues. If duplicate data is a copy of a row, redundant data is a copy of a column. However, unlike duplicate data, the columns do not have to be perfect copies to be redundant—they just need to indicate the same thing. Let's look at the following example table:

ID	Sex	Male	Female
84927	M	TRUE	FALSE
69427	M	TRUE	FALSE
10374	F	FALSE	TRUE
58264	M	TRUE	FALSE
90162	F	FALSE	TRUE

Figure 4.2 – Redundant data

In *Figure 4.2*, we see four columns with unique names, and each row is unique, so what is redundant? The Sex, Male, and Female columns all give the exact same information. More to the point, you could use any one of these columns to predict the other two with 100% accuracy. If the Sex value is M, then Male will always be TRUE and Female will always be FALSE, and if the Sex value is F, then Male will always be FALSE and Female will always be TRUE.

We now know that redundant data is made up of columns that can be used to perfectly predict other columns, but why is that a problem? Sitting in a dataset, there is no problem, though it is just clutter taking up extra space. However, the moment you put it into a statistical model, you get what is called **multicollinearity**. Multicollinearity in itself is just multiple independent variables in a model that are highly correlated or can predict each other. Specific ramifications of multicollinearity get very academic very quickly and are the subject of some debate, but it can make results harder to interpret because you don't get a clear picture of which independent variable has which impact, and while it may run accurately on a sample, that model does not translate well when applied to the entire population. Long story short, it mucks things up.

Important note

For more information on this topic, look for guides on multicollinearity and overfitting.

A decent article on multicollinearity can be found here:

`https://towardsdatascience.com/multi-collinearity-in-regression-fe7a2c1467ea`

A quick guide for overfitting can be found here:

`https://www.ibm.com/cloud/learn/overfitting`

Just know that there is some debate, and a lot of popular information is coming from *Wikipedia*, so take anything you read with a grain of salt.

Now that we have a rough idea and what it is and why it is a problem, let's discuss what to do about it. The safest thing to do is to delete redundant columns because they can't mess with your models if they don't exist and they will no longer take up space. Having an extra column may not matter if you only have a few thousand rows, but once your dataset gets millions or even billions of rows, those extra columns take up a lot of memory. That said, sometimes you don't have the option of deleting them; maybe they are there for a reason, or maybe you just don't have permission. In these cases, the answer is simply to create a subset of your data that does not include the redundant columns and use the subset to create and run statistical models.

Let's look at a quick example of creating a subset in *Figure 4.3*:

```
MyData.head()
```

	ID	Sex	Male	Female
0	84930	M	True	False
1	75982	F	False	True
2	19038	F	False	True
3	48902	M	True	False
4	10948	F	False	True

```
MySubset = MyData.drop(["Male","Female"], axis=1)
MySubset.head()
```

	ID	Sex
0	84930	M
1	75982	F
2	19038	F
3	48902	M
4	10948	F

Figure 4.3 – Creating a subset

In this example, we start with our data saved in a variable called `MyData`. We use the `head()` function to see what is in it and find a dataset similar to the data we saw in *Figure 4.2*. As you recall, this has redundant data. To create a subset, we just define a new variable called `MySubset`, and we fill it with the data from `MyData`. However, we use the `drop()` function to exclude the redundant variables. When we use the `head()` function to look at `MySubset`, we see that it only includes the columns we want. We can then run analyses on `MySubset` without having to worry about multicollinearity. In fact, it is generally good practice to work on a copy of your data instead of the original, so this is not a bad practice even if you do not need a subset. If you accidentally delete a variable you need in a copy, you can just make a new copy, but things get trickier when you go around deleting things in your original data. Note that, even if you are using Python, there are many different ways to create a subset—this is just one approach.

We have covered what to do about unwanted extra data, but often you will have the opposite issue, where random chunks of your dataset are just blank. In the next section, we will discuss options for dealing with missing data.

Dealing with missing data

Missing or incomplete data is a problem every data analyst will have to face at one time or another. Data can be missing for any number of reasons. Maybe someone just didn't enter the data, maybe it's a survey and the person didn't answer the question, or a measurement couldn't be taken for whatever reason. No matter the reason, holes in your dataset happen all the time, and it is something that needs to be addressed.

From a data analytics point of view, the biggest problem is that most analyses won't run with null values in the data. You get an error message, and you can't run the code until you have done something about all the gaps. From a statistical point of view, it is a little more complicated. Removing data reduces the statistical power of the analysis, and it can even drop the number of observations below what is required for a specific analysis. Perhaps the biggest problem is that sometimes what is missing is significant in and of itself, so how you deal with it can cause bias.

> **Important note**
>
> How to handle missing data is a hot-button topic, and there is a lot of debate around it. There are a lot of people who argue for one specific method to use on all types of missing data, no matter what. The fact of the matter is, despite popular belief, there is no one perfect approach. Every method to deal with missing data has some flaws and the potential to introduce some bias. Often, a company will already have a policy for dealing with missing data, and if it doesn't, pick a method that makes you comfortable.

Before we go any further, let's discuss the different types of missing data.

Types of missing data

Missing data can be broken down into three main categories. These groups are determined by why the data is missing. The reason why knowing the category of your missing data is important as it lets you know how much this missing data will influence your outcome, or how much it will bias your results. Generally speaking, the more random everything is, the less bias you will have. If all of your missing data is completely random, you will have little to no bias. If all of your missing data is not random and is missing for a specific reason, then you have a high probability of introducing bias.

Missing Completely at Random (MCAR)

If the data is MCAR, it means that you cannot find any connection between the missing values and the values that are still present. This data is not intentionally omitted, but just due to random chance. This is the best kind of missing data. Yes—it still reduces the amount of data, but it will not add any inherent bias. Let's look at an example:

Employee ID	LastName	FirstName	Department	Years With Company
83784	Benhill	Floyd	Sales	
64986	Chane	Jill	1T	1
93671		Richard	HR	15
37816	Smith	Trudy	Sales	21
73891	Doe	John	IT	2
20179	Brown	Olivia		18
	Crow	Steven	Sales	14
74982	Burns	Charles	Sales	22

Figure 4.4 – MCAR

In *Figure 4.4*, we see a dataset that shows common MCAR properties. The data that is missing is random and seems scattered throughout all variables.

Missing at Random (MAR)

MAR is another type of missing data, and its name is slightly confusing because it is not exactly random. The idea is that the missing data is connected to another variable that has been recorded. This means that it is not completely random, and that missing data may introduce some bias. Here is an example:

Employee ID	LastName	FirstName	Department	Years With Company
83784	Benhill	Floyd	Sales	12
64986	Chane	Jill	IT	
93671	Hanson	Richard	HR	15
37816	Smith	Trudy	Sales	21
73891	Doe	John	IT	
20179	Brown	Olivia	H R	18
80781	Crow	Steven	Sales	14
74982	Burns	Charles	Sales	22

Figure 4.5 – MAR

Figure 4.5 is an example of MAR. We see there is data missing, and it is all in the same column, or variable—YearsWithCompany. Here, we can see that there is a connection to another variable; in this case, Department. In all of the cases where the YearsWithCompany value is missing, the Department variable is equal to IT. Right now, we don't know why, but we know that if Department is IT then we are more likely to be missing the value of YearsWithCompany. This example is simplified, and all of the values are missing when the condition is met, but it might just be a significant portion. 45% of men over the age of 45 may choose not to answer a question on a survey, for example. The basic idea that the missing data is caused by one of the variables that you have recorded is what you need to know.

Missing Not at Random (MNAR)

Finally, the third type of missing data is MNAR. This type of data has a connection to some variable or type of information that was not recorded. There was some influential factor that is unknown to you that stopped this data from being entered into your database. This data is the most problematic and the most likely to cause bias in your results. Here is an example:

Employee ID	LastName	FirstName	Department	Years With Company
83784	Benhill	Floyd	Sales	12
64986	Chane	Jill	IT	1
93671	Hanson	Richard	HR	15
37816	Smith	Trudy	Sales	21
	Doe	John	IT	2
	Brown	Olivia	HR	18
	Crow	Steven	Sales	14
	Burns	Charles	Sales	22

Figure 4.6 – MNAR

In *Figure 4.6*, we see that the Employee ID column has half of the values missing. The missing data is grouped together in the same variable, and it is very specific, but there is nothing the missing values have in common with the other variables. There is no trend in any of the other columns that would explain the missing data. Here, it is safe to assume that there is some reason for the missing data that is not recorded in your dataset. This one requires more investigation.

Deletion

Deletion is the easiest and most used approach to missing data. Have holes in your dataset? Remove them. Done. Simple, right? This is a common approach, but it isn't perfect. For one thing, it reduces the amount of data that you have available to you. If your sample is already small, or if you are missing values in half of your records, it hurts to lose that much data.

> **Important note**
> *ALWAYS* work on a copy of your data. *NEVER* work on the original raw data. This is especially important when we are talking about the deletion of data. If you accidentally delete something from your original data, it is gone forever. That is bad.

Also, some people argue that you should only delete data if it is MCAR because you would otherwise risk introducing bias. Let's go over the different methods of deleting missing values.

Listwise deletion

Listwise deletion, occasionally called **casewise deletion**, is considered the more drastic approach. It means that if a single value is missing from an observation, you delete the entire thing. Across the board, an entire row is deleted when this happens. The entire case disappears, you are checking it off the list, and this row is just gone. This approach is popular because it is clean and efficient. If the missing data is MCAR and you have a huge dataset, there is no harm in just dropping huge chunks of data that you don't need.

Pairwise deletion

Pairwise deletion is a little more strategic. This is where you are only deleting or omitting specific values that are missing. How does this work? To put it simply, you are trying to avoid throwing out good data with the bad. You will probably run analyses that don't use the value that is missing, which means you can still use the rest of the data in the row. When it comes to an analysis that requires a variable that is missing, then you delete it. This method still has the potential to introduce bias with the missing values are not MCAR, but it is more conservative, so it is a popular approach for those who don't have so much data that they can throw away large chunks without blinking. You keep more data, but it can also be more complicated to actually perform.

Variable deletion

Variable deletion comes across as a little extreme. To remove an entire column means that you will not be able to use that variable at all for anything. That said, if you are missing over half of the values for a specific variable, it may make sense to drop it. Usually, at that point, you do not have enough values to run an analysis accurately anyway. More importantly, there is almost a guarantee that there is some mysterious reason that much data is missing and that you will introduce a large amount of bias into the results by trying to use it. If you know using that data will cause trouble, then you may be better off deleting the entire variable and not using it.

Filtering

Filtering is listed with the deletion methods because it is effectively the same thing. You are removing values from a dataset to create a subset of data that has no missing data. You can use filters instead of deletion for listwise or pairwise approaches. The advantage of using filters instead is simply that you can remove a filter to have access to that data again. All of the same rules and limitations of deletion still apply to filtering.

Imputation

Imputation is the opposite of deletion. Instead of removing data with holes, you just try to fill those holes. What you try to plug in those holes is what differentiates the different kinds of imputation. Now, there are several kinds of imputation, but we will only cover the most common here.

Mean, median, or mode imputation

We will discuss specifically what mean, median, and mode are later in this book, as well as how to calculate them. For now, you just need to know that they are different ways to estimate the middle of a dataset. This means that you are guessing what the most common number is and using it to fill in the gaps. Because this value is common for the rest of the dataset, it means that there is a decent chance that it is close to what those values would have been anyway. The main problem with this approach is that it inherently introduces bias, by skewing your data toward the middle. To put it another way, it artificially normalizes your data. Some of the values that you are replacing with a value from the middle may have been high or low, and there is no way to know.

Now, a perk of this approach is that it can also be used on MAR data as well as MCAR data. Since you know what is influencing the missing values, you can control it. For example, if most of the missing values come from a specific age range, you can find the mean value of people specifically in that age range as opposed to the mean of all of the data points.

Hot deck imputation

I will start by saying that this approach is not as popular as it used to be. This is largely because it is more complicated. There are several variations to this approach, but the general idea is simple. You take random values from elsewhere in your dataset and use them to plug in the holes. There are ways to adjust for which values are selected, which values can go in which holes, and even how many times a value can be used to fill holes. The idea behind this approach is that it is supposed to be more random, so it should be less likely to introduce bias.

Even this method has issues, and several of those issues have different variations as solutions, but it is a rabbit hole, and the deeper you go, the more complicated and theoretical it gets. That said, this method can account for MCAR and MAR data in the same way other imputation methods can.

Now, let's move on to something a little more advanced.

Interpolation

Interpolation is similar to imputation because it is still about plugging holes. Sometimes, interpolation is even considered a kind of imputation. There is an important difference, though. Where imputation takes values from the dataset as a whole and tries to plug in the holes, interpolation tries to estimate the specific values for every hole, using the other values for that data point to calculate what it should be and using the rest of the dataset as a reference.

There are several different kinds of interpolation, including regression interpolation (sometimes called linear interpolation, or **K-Nearest Neighbors** (**KNN**) interpolation), but these are all actually analytical modeling approaches. You are running an analysis on the data to fill the holes in the data so that you can use the data to run analyses.

Even interpolation is not perfect and can introduce bias. These methods get very technical very quickly, but how accurate they are depends heavily on the data that is there, just as with any other statistical model or **machine learning** (**ML**) algorithm. These methods should automatically adjust for MAR and are fine with MCAR.

Dealing with MNAR

You will notice that none of these methods can safely account for MNAR data without introducing bias. MNAR is problematic. There is obviously a reason why the data is missing, but you don't know what it is. Different people approach MNAR in different ways. Some people try to guess what the influencing variable is, then collect data again, this time looking for that one variable. However, starting the data collection process over is not always an option. Some people say to heck with it and treat it like MCAR, deleting or imputing as they normally would, and accepting that some bias will be introduced, because they are missing some key piece of data. The most conservative approach is to not use the data from that variable. If you know data is missing from that variable and you don't know why, not using the entire variable is the only way to avoid introducing bias.

Here, we can neatly differentiate between this kind of missing data and that kind, but in a large set, you may very well have a combination of different types of missing data or even all three. Unless you want to accept the risk of using one method on every type of missing data, you will need to look at the data that is missing carefully, try to figure out which kind each is, and deal with it using an appropriate approach.

However, extra data and missing data are not the only things that can go wrong and make a dataset messy. What if the values are not what you would expect?

Understanding invalid data, specification mismatch, and data type validation

Sometimes mistakes are made, and the data is just wrong. Especially when data is entered manually, there will be typos or values put in the wrong cells. Even if the person entering the data has 99% accuracy, that still means one mistake for every hundred values entered, and when you have millions of data points, each made up of dozens of variables, those mistakes add up. Here are a few of the more common mistakes.

Invalid data

Invalid data happens when data does not match expected values or ranges. This is usually caused by a typo but is often just a matter of format. Let's look at an example:

City
los angeles
LA
Los Angeles
Los Angelus
la
los angeles

Figure 4.7 – Invalid data

In *Figure 4.7*, we see a simple variable called `City`. Every value in this column is talking about the same city, but to software or a programming language, this is talking about six different cities. Spelling, capitalization, and abbreviations can all create something recognized as a unique value. What's that?— you say. The first value and the last value are exactly the same? What you don't see, what you can't see, is the invisible space that comes after the last value, making it `los angeles ` instead of `los angeles`. Yes—even spacing can create invalid data.

Needless to say, this is a problem. If you are treating the same city as six different cities, you are going to get some weird results. There are all kinds of creative tricks for dealing with invalid data, such as making all of the entries uppercase or lowercase, which removes any issues from capitalization. If you can, having the data entered from a drop-down menu helps reduce invalid data, because people can only choose valid options. There are even tricks for removing spaces.

The most important thing you need to learn how to do is to identify invalid data. There are different approaches for different software and programming languages, but the easiest approach is to generate a list of unique values, or even generate a frequency table. If you see more unique values than you should, then chances are you have some invalid data. Go through the list of unique values until you find the values that shouldn't be there, then replace them with the correct values.

Specification mismatch

Specification mismatch is similar to invalid data, but this is where the value does not make any sense in that variable. Here, we are specifically talking about the value having a different data type than the other values in a variable. Let's look at a quick example:

Cost Per Click
$1.82
$0.95
Toast
$2.10
$1.39

Figure 4.8 – Specification mismatch

In *Figure 4.8*, we see a variable that is looking at cost per click, a common metric in marketing that looks at how much money was spent to get someone to click on a link or visit a website. One of these values is not like the others. In a variable that is formatted as a currency data type, there is a string of letters: Toast.

This is an issue. Having even a single value in a variable be the wrong data type will cause issues. When you try to run an analysis using this variable, it will expect every value to be the same data type, so when it comes to something that is not what it expects, it causes an error. This means Toast will crash any analysis you try to run on that column.

On a side note, a specification mismatch can also happen when the entire variable is not the data type that the analysis requires or expects. For example, if you have a variable that lists a car's manufacturer and you try to find the average, you will get an error. The equation to calculate the average is expecting numbers, not brand names.

Fixing this comes in a few different forms. If a single value is the wrong data type, you can just fix or remove that one value. If the variable is the wrong data type, you can either change the analysis you are trying to run or change the data type of the variable. Alternatively, you can avoid the problem altogether.

Data type validation

Data type validation is the process of manually or automatically checking the data type of a variable to avoid specification mismatches. There are different ways to do this. The best solution is to make it so that when the data is entered, the correct data format is required. When you are filling out an automated form and you put your name where they want your phone number, chances are you will get an error message stating that what you put in is an invalid entry. This is data type validation.

Such a simple solution is not always available to you, so sometimes you have to manually check the data type of each variable, to make sure it is what you, and your analyses, expect it to be.

Sometimes the issue is not with the data types, but with how the data is distributed.

Understanding non-parametric data

Non-parametric data is data that does not follow a normal or well-known distribution. Let's briefly cover what that means. To put it simply, if you visualize the frequency of a numerical variable, you will often have the most common values higher in the middle. As you move away from the middle, the frequency decreases as you get larger or smaller. This creates what is commonly known as a bell curve or a normal distribution, as depicted in *Figure 4.9*:

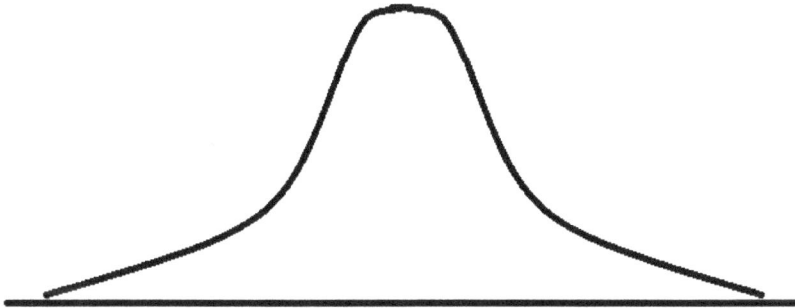

Figure 4.9 – Normal distribution

In statistics, a distribution can be used to predict the probability that a new value will be any specific number. We will talk more about distributions later in this book. For now, know that there are a few distributions that are common and well-understood, such as the normal distribution depicted in the preceding screenshot.

Non-parametric data does not follow any of the common distributions. To put it another way, the data is in such a funny shape that it doesn't even come close to any of the common shapes, like the one seen here in *Figure 4.10*:

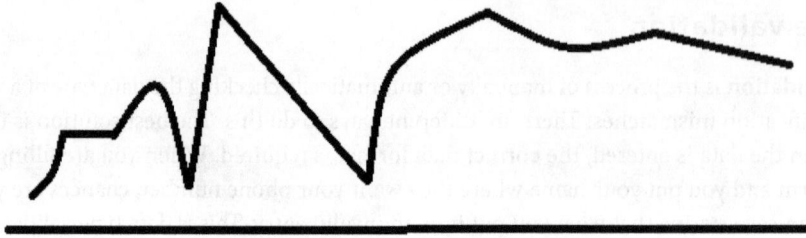

Figure 4.10 – Non-parametric distribution

Sometimes, data just comes in a weird shape. On occasion, you will be able to massage the data into something closer to a known distribution, but often, you will just have to accept that it is non-parametric. Okay—so your data is non-parametric. Why is that a problem?

The issue is that the majority of common statistical analyses are inherently parametric. These parametric analyses assume that your data is in a specific distribution, and if it isn't, you will not get accurate results. This means that if you have a non-parametric variable, you need to look into specific non-parametric statistical analyses, sometimes called distribution-free tests because it doesn't matter what the distribution is for these tests.

Next, let's talk about specific data points that are so far away from the middle that a single point can skew your entire dataset.

Finding outliers

Outliers represent data points that are so much larger or smaller than the rest of your data points that they cause issues. They can even skew your entire distribution. That one point can throw off all of your results. Here's an example:

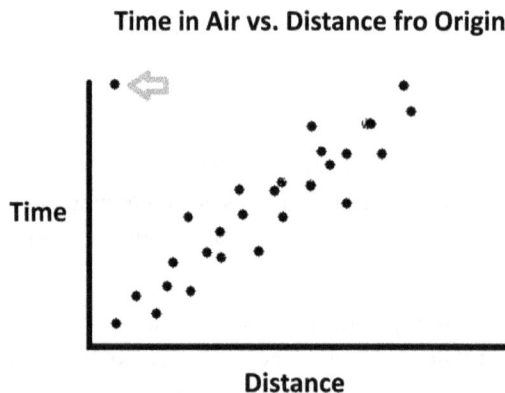

Time in Air vs. Distance fro Origin

Figure 4.11 – Outlier

In *Figure 4.11*, we see a scatter chart with the majority of the data points together, but one single point in the upper left-hand corner by itself. Somehow, this observation spent a lot of time in the air but traveled almost no distance. That lone point is probably an outlier. If you were to include that outlier in your calculations, it would artificially pull all of your numbers up and to the right.

In the academic field of statistics, outliers are another controversial issue. There are arguments on how to define what is an outlier and there are arguments about what to do with them. There are specific statistical analyses that only say whether or not a single point counts as an outlier. Each and every potential outlier is considered individually and investigated.

Luckily, data analytics as a field generally takes a more pragmatic approach. It creates a range, and anything outside of that range is considered an outlier automatically and gets deleted. The end.

There are several ways to find the range of acceptable data points, but there are two common approaches. You can calculate the standard deviation, which we will explain later in this book, and anything that is more than three standard deviations away from the mean is an outlier and gets deleted. The other common approach is to calculate the **interquartile range** (**IQR**), using a cutoff of 1.5 IQR.

Much like dealing with missing data, there is no perfect approach to dealing with outliers. If your company already has a policy, stick with that. If it doesn't, find a method that makes you comfortable, though you should probably avoid just eyeballing it and guessing.

> **Important note**
>
> A great resource for hands-on practice with outlier detection can be found here:
>
> `https://analyticsindiamag.com/a-complete-guide-to-outlier-detection-with-hands-on-implementation-for-beginners`

This chapter has covered how to clean data by accounting for things that make the data impossible to run or will cause issues with the results. Cleaning your data is an important first step to making it usable because it removes things that will cause errors. In the next chapter, we will talk about data wrangling and manipulation, or the process of getting the data in a specific shape so that you can use it for a specific analysis.

Summary

This chapter covered how to clean and process data. First, we covered the difference between duplicate data and redundant data and how to deal with each. Then, we talked about the highly debated question of what to do with missing data, which covered the different types of missing data, different methods of deleting missing data, different types of imputation, and interpolation. Next, we went over common issues such as invalid data, specification mismatch, and data type validation. Then, we covered non-parametric data, what it is, and what that means for you. Finally, we discussed outliers and how to address them. This wraps up how to clean your data. In the next chapter, we will cover how to wrangle your data and get it into a shape you can use!

Practice questions

Let's try to practice the material in this chapter with a few example questions.

Questions

1. In regard to the following table, you can state:

ID	Sex	Male	Female
84927	M	TRUE	FALSE
69427	M	TRUE	FALSE
69427	M	TRUE	FALSE
10374	F	FALSE	TRUE
58264	M	TRUE	FALSE
90162	F	FALSE	TRUE

   A. This table has redundant data

   B. This table has duplicate data

   C. This table has redundant data and duplicate data

   D. This table has neither redundant data nor duplicate data

2. When dealing with missing data, which of the following is *not* a type of deletion?

   A. Listwise

   B. Singlewise

   C. Pairwise

   D. Variable

3. Which types of errors can be found in the following table?

Employee ID	LastName	FirstName	Department	Years With Company
83784	Benhill	Floyd	Sales	12
64986	Chane	Jill	IT	1
93671	Hanson	Richard	HR	15
37816	Smith	Trudy	Sale	21
73891	Doe	John	Information Technology	11

A. Invalid data

B. Specification mismatch

C. Data type validation

D. None of these

4. Non-parametric data is a problem because:

A. It has a normal distribution

B. You can't use parametric analyses on it

C. It has no distribution

D. There is no problem with non-parametric data

5. When in a dataset on the weight of human babies at birth, you come across a value of 8,000 lb. What can you deduce?

A. It is probably an outlier, and you should check your ranges to be sure

B. The data point will probably skew your results

C. It is probably safe to delete this data point

D. All of these

Answers

Now, we will briefly go over the answers to the questions. If you got one wrong, make sure to review the topic in this chapter before continuing:

1. The answer is: *This table has redundant data and duplicate data*

    The data in the second row is repeated in the third, meaning there is duplicate data, and the Sex, Male, and Female columns are redundant.

2. The answer is: *Singlewise*

    Listwise deletion, pairwise deletion, and variable deletion are all types of deletion covered in this chapter.

3. The answer is: *Invalid data*

    Here, there is a typo that differentiates the Sales department from the Sale department. Also, Information Technology is spelled out in one place and abbreviated in another. As is, this would be read as five unique departments instead of three.

4. The answer is: *You can't use parametric analyses on it*

    Non-parametric data requires non-parametric statistical analysis.

5. The answer is: *All of these*

    4-ton human babies are not likely. You should check to make sure it is an outlier, then delete it, or else; otherwise, it will skew all of your results.

5

Data Wrangling and Manipulation

It is not an exaggeration to say that the majority of work done by the average data analyst revolves around preparing data for use. A large part of this is cleaning the data, as covered in the previous chapter, but it is more than just dealing with things that will cause errors or introduce bias. You will often have to get the data into a specific shape or format before you can use it. This step is often called **data wrangling** or **manipulation**. To be clear, when we use the word "manipulation," we do not mean we are changing the outcome in any way; we are using it in the literal sense of handling and managing the data in a skillful way.

In this chapter, we will go over some of the most important skills in the data-wrangling process. We will talk about the different methods to combine datasets including different types of joins, blends, concatenation, and appending. Next, we will look at the addition of variables that add meaning, such as derived or reduced variables. Then, we will cover breaking down chunks of text into something usable with parsing. Then, we will discuss recoding variables. Finally, we will go over some common tips and tricks to help you shape your data.

In this chapter, we're going to cover the following main topics:

- Merging data
- Calculating derived and reduced variables
- Parsing your data
- Recoding variables
- Shaping data with common functions

Merging data

We will be covering all of the different ways to combine datasets. While it is nice when every variable you need is conveniently stored in one table or dataset, it doesn't always happen. A large database will have several different tables and sometimes you need to compare two variables from two different tables, or you have part of a table stored in one part of the database and another month or year of the same table stored in another part of the database. There are many different ways to merge data, depending not only on the software or programing language you are using but also on how you need the tables to come together.

Key variables

Before we can jump in and talk about merging tables, we need to understand the concept of key variables. A key variable is simply a variable that is in both tables being merged, which allows all of the rows to be matched up in a way that makes sense.

Customer Names

FirstName	LastName
Laurence	Smith
Betty	Brown
Phil	Hook
Jen	Roark
Jona	Cox

Customer Locations

City	State
Austin	TX
Denver	CO
Tulsa	OK
Phoenix	AZ
Seattle	WA

Figure 5.1 – Customer names and customer locations

As you can see in *Figure 5.1*, we have two tables. The table on the left contains names and the table on the right contains locations. However, we have no way of knowing which location is connected to which name. Let's look at another example.

Customer Names

CustomerID	FirstName	LastName
1	Laurence	Smith
2	Betty	Brown
3	Phil	Hook
4	Jen	Roark
5	Jona	Cox

Customer Locations

CustomerID	City	State
3	Austin	TX
2	Denver	CO
5	Tulsa	OK
1	Phoenix	AZ
4	Seattle	WA

Figure 5.2 – Customer names and customer locations with key variable

In *Figure 5.2*, we now have a key variable: CustomerID. Because they now share this variable, we can see how the data in the table on the left is related to the table on the right. Trying to join tables without a key variable is problematic at best, so try to plan ahead and make sure your database is structured in such a way that this is not an issue. This is why database schemas such as star or snowflake have a central table, called a **Key Table**, that is nothing but key variables, and you need to make sure that every other table has at least one key variable. Let's see what this looks like.

Customer Names

CustomerID	FirstName	LastName
1	Laurence	Smith
2	Betty	Brown
3	Phil	Hook
4	Jen	Roark
5	Jona	Cox

Key Table

CustomerID	LocationID
1	4
2	2
3	1
4	5
5	3

Customer Locations

LocationID	City	State
1	Austin	TX
2	Denver	CO
3	Tulsa	OK
4	Phoenix	AZ
5	Seattle	WA

Figure 5.3 – Customer names and customer locations with Key Table

In *Figure 5.3*, we see this in action. Even if the tables cannot be merged together directly, they can both be merged with the Key Table, so they can still be connected. Using a schema like this, you will be able to merge any tables within the database.

Joining

A join is sometimes called a **merge** depending on the program you are using, but you can consider a merge as the general concept of bringing data together, whereas a join is a specific kind of merge that creates a new table from two different tables. That said, the majority of merges are joins, so the terms are sometimes used interchangeably.

There are a few different kinds of joins on the exam:

- Inner joins
- Outer joins
- Left joins
- Right joins

These different types describe how the tables will relate to one another. There are other types for fringe cases, but these are the most common.

Inner joins

An **inner join** is a join where the new table only includes values that are found in both of the old tables. This is often visualized as a Venn diagram, as seen in *Figure 5.4*.

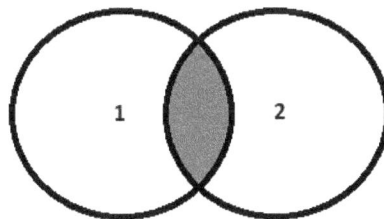

Figure 5.4 – Inner join Venn diagram

This is easy enough to say, but it helps to have an actual example. Let's see what this looks like.

ClientID	Name		ClientID	Name	City		ClientID	City
Left Table			**Joined Table**				**Right Table**	
1	Smith, Laurence		1	Smith, Laurence	Austin, TX		1	Austin, TX
2	Brown, Betty		2	Brown, Betty	Denver, CO		2	Denver, CO
3	Hook, Phil		3	Hook, Phil	Tulsa, OK		3	Tulsa, OK
4	Roark, Jen						7	Phoenix, AZ
5	Cox, Jona						8	Seattle, WA
6	Humbert, Ren						9	Baltimore, MD

Figure 5.5 – Inner join tables

In *Figure 5.5*, we have an example that looks at joining a table of client names and client locations. The table on the left contains the client names, the table on the right contains the client locations, and the table in the middle shows the table that would be created if the other tables were merged with an inner join. We can see that there is information in the client name table that is not in the client location table and vice versa. As we can see, with an inner join, every data point that does not exist in both tables is dropped in the new table.

Inner joins inherently drop most data, so they often result in a smaller final table. This approach is considered the most conservative and leaves you with the cleanest final dataset.

Outer joins

Outer joins, sometimes called full joins, include every data point, whether or not it is in both tables, as seen in *Figure 5.6*.

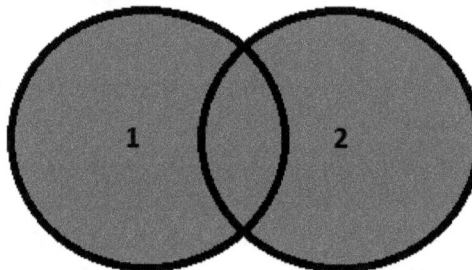

Figure 5.6 – Outer join Venn diagram

In other words, every data point in both tables is present in a full join. Let's look at an example.

ClientID	Name		ClientID	Name	City		ClientID	City
Left Table			**Joined Table**				**Right Table**	
1	Smith, Laurence		1	Smith, Laurence	Austin, TX		1	Austin, TX
2	Brown, Betty		2	Brown, Betty	Denver, CO		2	Denver, CO
3	Hook, Phil		3	Hook, Phil	Tulsa, OK		3	Tulsa, OK
4	Roark, Jen		4	Roark, Jen	NULL		7	Phoenix, AZ
5	Cox, Jona		5	Cox, Jona	NULL		8	Seattle, WA
6	Humbert, Ren		6	Humbert, Ren	NULL		9	Baltimore, MD
			7	NULL	Phoenix, AZ			
			8	NULL	Seattle, WA			
			9	NULL	Baltimore, MD			

Figure 5.7 – Outer join tables

In *Figure 5.7*, we can see an example of an outer join. The tables on the left and right are the same as in the previous example, but the one in the middle is the result of an outer join. Every data point is represented here. This is the most inclusive approach and creates the largest tables, but, as you can see, it will also have the most null values. These null values are created when you bring in a data point from one table that has no value in the other table. This approach is popular when you have a small sample and need to use every data point you have. Outer joins are sometimes used in conjunction with pairwise deletion to get the most out of a small dataset.

Left joins

Left joins contain all the data points in the left table and only values in the right table that have a match in the left table, as seen in *Figure 5.8*.

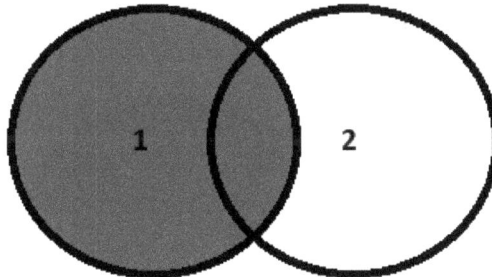

Figure 5.8 – Left join Venn diagram

Before we do anything else, let's clear up what "left" and "right" mean in this context. Often when you are preparing for a join, you select one table and then a second table that you want to merge. The table that you select first is considered the "left" table, and the table you select second is the "right" table. Alternatively, some programming languages will have you specify which table is considered "left" or "right." Yet another way will have you select a default table, which will become "left," and a table you want to add to it, which will become "right."

No matter what software or programming language you are using, make sure you have it clear in your mind which table is considered "left" and which table is considered "right." Let's look at an example of this.

Left Table			Joined Table				Right Table	
ClientID	Name		ClientID	Name	City		ClientID	City
1	Smith, Laurence		1	Smith, Laurence	Austin, TX		1	Austin, TX
2	Brown, Betty		2	Brown, Betty	Denver, CO		2	Denver, CO
3	Hook, Phil		3	Hook, Phil	Tulsa, OK		3	Tulsa, OK
4	Roark, Jen		4	Roark, Jen	NULL		7	Phoenix, AZ
5	Cox, Jona		5	Cox, Jona	NULL		8	Seattle, WA
6	Humbert, Ren		6	Humbert, Ren	NULL		9	Baltimore, MD

Figure 5.9 – Left join tables

In *Figure 5.9*, we can see the same example we have been using, but now the table in the middle represents a left join. In this case, the client name table is "left" and the client location table is "right." As you can see, every client name is listed, but only the client locations that have names are included. You will have more null values than an inner join but fewer than an outer join. Left joins are less about the amount of data and much more about priority. The data points in the left table are more important, so you want to make sure you have all of them, no matter what. The information in the right table is only there to augment what is in the left table.

Right joins

Right joins are exactly what you imagine. They are the exact opposite of a left join; you take all data points in the right table and only the data points in the left table that have a match in the right, as seen in *Figure 5.10*.

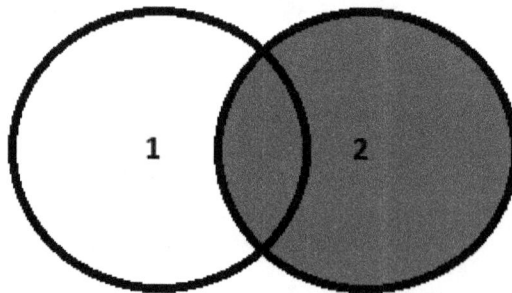

Figure 5.10 – Right join Venn diagram

If you understand left joins, then you understand right joins. Let's see what this looks like.

| **Left Table** | | **Joined Table** | | | **Right Table** | |
ClientID	Name	ClientID	Name	City	ClientID	City
1	Smith, Laurence	1	Smith, Laurence	Austin, TX	1	Austin, TX
2	Brown, Betty	2	Brown, Betty	Denver, CO	2	Denver, CO
3	Hook, Phil	3	Hook, Phil	Tulsa, OK	3	Tulsa, OK
4	Roark, Jen	7	NULL	Phoenix, AZ	7	Phoenix, AZ
5	Cox, Jona	8	NULL	Seattle, WA	8	Seattle, WA
6	Humbert, Ren	9	NULL	Baltimore, MD	9	Baltimore, MD

Figure 5.11 – Right join tables

In *Figure 5.11*, we look at the same example but with the center table representing a right join. The client name table is still left and the client location table is still right. Every data point from the client location table is represented, and only the data points in the client name table that have a corresponding value in the client location table are present.

There are some data analysts who use both left joins and right joins equally as the occasion calls for it, but there are just as many who only do left joins or right joins and just change which table is considered "left" or "right." Either way doesn't really matter; they do the exact same thing. What does matter is being able to identify which table is left, which table is right, and what will happen if you are asked to perform a specific kind of join.

Blending

Blending is another type of merging and it is relatively new. It has only come about recently, because of Tableau, but it has been included in a few other programs since. There are several articles that list every exact difference between blends and joins, but a lot of them are academic. Data blending temporarily links information in one table to information in another through what amounts to a left join. That said, you aren't actually creating a new table, just a one-way connection between two tables, so you can use the variables of both in a given software based on a key variable. There are lots of little differences, such as when things are merged versus when they are aggregated, but, at a high level, it doesn't make a huge difference. What you need to know is that blending is like joining, but it creates a temporary one-way link between two tables.

Concatenation and appending

Concatenation and appending are related concepts, with slightly different executions. Concatenation is merging things together in a series while appending is adding a new value to the end of an existing series. Sound similar? They are.

The term concatenate is kind of a general term that can mean a few different things in practice. It can be used in a general programming sense to take multiple short text variables and combine them to make a single long text variable. However, in data analytics, concatenation is usually in reference to adding a group of additional values to a dataset. Whereas joining adds on additional columns, concatenation adds on additional rows. This will look something like *Figure 5.12*.

May Table				Concatenated Table				June Table		
Date	QuantitySold	Income		Date	QuantitySold	Income		Date	QuantitySold	Income
5/1/2022	4	759.96		5/1/2022	4	759.96		6/5/2022	8	1519.92
5/8/2022	3	569.97		5/8/2022	3	569.97		6/12/2022	6	1139.94
5/15/2022	7	1329.93		5/15/2022	7	1329.93		6/19/2022	4	759.96
5/22/2022	2	379.98		5/22/2022	2	379.98		6/26/2022	7	1329.93
5/29/2022	6	1139.94		5/29/2022	6	1139.94				
				6/5/2022	8	1519.92				
				6/12/2022	6	1139.94				
				6/19/2022	4	759.96				
				6/26/2022	7	1329.93				

Figure 5.12 – Concatenated table

As you can see in *Figure 5.12*, we have three tables. The tables on the outside represent a different month's worth of data, collected weekly, and the table in the middle is the concatenated table. That said, they are all looking at exactly the same variables. Concatenation stacks these on top of each other to make one large table.

Appending does not merge large datasets; instead, it adds a single value to the end of a series that already exists. This can be as simple as adding a new item to the end of your shopping list.

Now that we have covered the most common merging methods, it is time to look at the creation of new variables that can give us more insight into our data quickly and easily.

Calculating derived and reduced variables

In this section, we will talk about specialized variables that you can create that will help you as a data analyst. You will find that raw data, even clean raw data, can be difficult to interpret. When looking at a functional dataset that is actively being used by a data analyst, you will almost always find variables that were not present when the data was originally recorded. Instead, these variables were added later and contain some logic that allows them to create a new value based on those that were recorded.

Derived variables

Variables that use logic that relies on other variables are broadly called **derived variables**, though some also refer to them as calculated variables or fields. The idea is just that this variable was not observed but was generated based on data that was observed. If this definition seems a little vague, that's because it is. There are as many derived variables as there are stars in the sky and they come in all shapes and sizes. The specific ones used in a dataset depend completely on what you need out of the data and change wildly from one field of analytics to another. The derived variables used by a marketing analyst look nothing like those used by a risk analyst. That said, derived variables largely fall into one of three categories: metrics, flags, or recodes. We will talk about recodes later in this chapter.

> **Important note**
>
> Remember, derived variables and the variables used to calculate them are often highly correlated and can be used to predict one another. This, if you remember from the last chapter, can cause multicollinearity. Try to avoid putting both into an analytical model if you can help it. Instead, use the observed variables or the derived variables.

Metrics are derived variables that calculate a number you can use to gauge the status of a data point. The majority of **Key Performance Indicators (KPIs)** are metrics. As the name would suggest, KPIs allow you to judge how something is performing and whether or not it is progressing toward a goal. Let's look at an example.

Date	Distance (m)	Time (s)	Speed (m/s)
1/1/2022	100	18.94	5.28
1/2/2022	100	16.43	6.09
1/3/2022	50	9.67	5.17
1/4/2022	200	33.8	5.92
1/5/2022	100	15.85	6.31

Figure 5.13 – Derived variable metrics table

In *Figure 5.13*, we see a simple table that is recording information about runs. The variables for Distance (m) and Time (s) are observed and recorded at the time of the run, but the Speed (m/s) variable is a derived variable. It is calculated by dividing Distance by Time. If you were training for a race, then the Speed variable is a KPI, because it is something that clearly shows how you are performing and you can track it to see if you are getting closer to or further from your goal.

If metrics are derived variables that show quantitative data, flags are derived variables that show qualitative data. Flags, sometimes called tags, are categorical variables that summarize the status of another variable or the data point as a whole. This kind of derived variable is useful in all kinds of ways but they are often binary variables that say whether or not a condition is met. Once you have a flag in place, it is easy to use it to filter your dataset or create conditional logic that will allow you to treat data points differently based on the flag variable. Let's see what this looks like.

CheckOutDate	BookID	ReturnDate	Date	DueDate	OverDue
3/14/2022	8473097	3/16/2022	3/22/2022	3/21/2022	
3/14/2022	8721809		3/22/2022	3/21/2022	OverDue
3/15/2022	3809893	3/19/2022	3/22/2022	3/22/2022	
3/16/2022	5837988	3/18/2022	3/22/2022	3/23/2022	
3/16/2022	4839801		3/22/2022	3/23/2022	

Figure 5.14 – Derived variable flags table

In *Figure 5.14*, we see a simple table from a library. `CheckOutDate`, `ReturnDate`, `Date`, and `DueDate` are observed variables that have been recorded. `OverDue` is a flag derived from a combination of the other variables. This way, instead of having to look at three or four different variables to figure out whether or not a book is overdue, you have one simple variable that you can use as an indicator.

Reduction variables

Reduction variables are a specific kind of derived variable that falls into the metric category. In data analytics, reduction variables, otherwise known as aggregate variables, are meant to reduce the volume of data. If you can summarize multiple variables, entire rows, or entire columns with a single variable or even a single number, you are reducing the sheer amount of data you have to handle. This becomes more and more important as datasets get larger and larger.

This has become such a problem when handling big data that some very advanced methods have been developed using machine learning algorithms, using decision trees or cluster analysis. Luckily, for the purpose of this exam, you only need to know the most basic methods of aggregation:

- Average
- Sum
- Maximum
- Minimum
- Count
- Distinct Count

These are exactly what you would expect. Instead of reporting an entire series of numbers, you only work with the average of those numbers, the total of those numbers added together, the largest number in the series, the smallest number in the series, the number of entries in a series, or the number of unique entries in a series, respectively. This process is pretty common. Some data analytics programs, such as Tableau, will aggregate every variable by default. Working with a smaller volume of data makes processing and analyzing data faster.

Reduction variables are all about bringing data together, but sometimes you receive data in a big chunk and you have to break it apart in order to use it.

Parsing your data

In this section, we will talk about what **parsing** is and some common ways it is used. Sometimes you will receive data in a format that is not readily usable. Whether you are pulling data from a website, working with JSON files, or have big chunks of text, you will need to parse your data. There are many different parsers that you can use, depending on what you need to parse, but the general idea is that you are breaking a single large piece of data into several smaller pieces of data that can be easily identified and processed.

Natural Language Processing (NLP) is a field of data analytics that specializes in analyzing, you guessed it, language. Spoken or written, NLP is trying to translate common speech into actionable data. Parsing is necessary for even basic NLP.

> **Important note**
>
> In reference to NLP, parsing is called tokenization because it is breaking up the text into words, and each becomes its own object or token.

Let's consider an example. We will use the sentence "This book makes me happy." Let's go ahead and store it in a variable called `Data` as follows:

```
Data = "This book makes me happy."
```

If you receive the preceding line as a piece of data, the computer accepts this as one whole and can do very little with it. However, we can use a parser on it like in the following line of code:

```
Data = ["This", "book", "makes", "me", "happy", "."]
```

Now, instead of a sentence, we have a list of words. We can remove words we don't need or punctuation, identify which part of speech each word is, or use sentiment analysis to determine what emotions are being conveyed. Here, the subject is `"book"` and the sentiment is `"happy"`, and there are no negatives, so even a computer can figure out that there is a positive sentiment toward `"book"`.

Even if we don't consider NLP, parsing is a powerful tool that can break down large chunks of code, identify what you want, store it neatly, and throw out what you don't need. It is a very effective way of changing the format of your data into something you can use, but it is not the only way. Sometimes, it is not about making your data smaller or bigger but simply changing how it is being communicated.

Now you know about parsing data, but what if your data is already in nice, neat chunks, but the chunks are the wrong data type? Let's talk about recoding.

Recoding variables

In this section, we will discuss all of the different ways you can recode variables to turn them into a format you can use. Recoded variables, as mentioned when we talked about derived variables, are a kind of derived variable. Instead of the focus being on summarizing data, the idea is to create a direct translation of a variable into a different format. This is how you turn quantitative variables into qualitative variables and vice versa. Recoding can try to make data easier to understand, but it is most often used because a certain analysis requires data to be in a specific format to run.

Recoding numbers into categories

When it comes to creating a category based on a number, ranges are the most common approach. Effectively, you are saying if the value of X is between A and B, it falls into this group, and if X is between B and C, it falls into this other group. Where specifically you make the cut-offs to set the ranges can get very complicated and is a matter of some debate. That said, as a data analyst, you often have a real-world reason for creating the groups that you are creating. Let's look at an example.

Date	Distance (m)	Time (s)	Speed (m/s)	SpeedCategory
1/1/2022	100	18.94	5.28	Slow
1/2/2022	100	16.43	6.09	Average
1/3/2022	50	9.67	5.17	Slow
1/4/2022	200	33.8	5.92	Slow
1/5/2022	100	15.85	6.31	Average

Figure 5.15 – Recode number to category

In *Figure 5.15*, we see the run data from before. However, we have added the `SpeedCategory` variable, which is a recode of the `Speed` variable. A quick glance at the `Speed` column will give you a number, but you may not know off the top of your head what that number implies. `SpeedCategory` has created groups based on the average speed of runners during the previous race. In this case, a range was created that included one standard deviation above or below the mean, which was the **Average** group. Anything above this range fell into the **Fast** group and another below this range fell into the **Slow** group. This will let you know at a glance whether a given speed was below the average, around the average, or above the average speed.

Recoding categories into numbers

Several analytical or machine learning models don't play well with text. Instead, categories are written as numbers. These are not a range but a direct translation. In other words, X = 1, Y = 2, Z = 3, and more. Let's see this in a table.

Month	UnitsSold	Color	ColorRecoded
August	432	Red	1
August	365	Blue	2
August	154	Yellow	3
September	398	Red	1
September	386	Blue	2
September	108	Yellow	3

Figure 5.16 – Recode category to number

In *Figure 5.16*, we see sales data about a specific product a company sells. This product comes in three different colors: Red, Blue, or Yellow. However, the analysis we want to run requires Color to be written as a number, so we have added the ColorRecoded variable, which translates the colors into numbers; where Red is 1, Blue is 2, and Yellow is 3.

Dummy coding

There is a particular kind of recoding called **dummy coding**. This technique is not as popular as it used to be, but it still has a few uses. There are certain analyses that require dummy coding, but it can also be used to give another layer of detail to an analysis. Dummy coding creates a new binary variable for every possible category in the original variable, where 1 indicates that this is the category that the original variable was and 0 for all other variables. This way, you can not only look at how much of an impact a variable has but also at how much of an impact each individual category within a variable has. The following table displays this nicely.

Month	UnitsSold	Color	Red	Blue	Yellow
August	432	Red	1	0	0
August	365	Blue	0	1	0
August	154	Yellow	0	0	1
September	398	Red	1	0	0
September	386	Blue	0	1	0
September	108	Yellow	0	0	1

Figure 5.17 – Recode dummy coding

In *Figure 5.17*, we see the same table as before, but now Color has been recoded using dummy coding, so we have Red, Blue, and Yellow as separate variables. You can do this manually, through logical functions, or some programs will have features that do this for you. Something important to note is that these will give you all of the variables, but you should never use all of the variables in the same model. First, obviously, don't use recoded variables in the same model as the original variable. Second, when using variables created by dummy coding, you always have to drop one of the new variables. It doesn't even matter which one, but if you have every dummy-coded variable in the model, they will perfectly predict one another and you will suffer from multicollinearity. Removing just one of the dummy-coded variables prevents them from predicting each other with 100% accuracy.

You are now an expert in recoding, but there are other ways to change data to make it more useful to you. Next, let's look at how to shape data with common functions.

Shaping data with common functions

In this section, we will talk about useful tools for shaping data that can be applied to almost any analytical program. There are lots of other tips and tricks that don't fit nicely in any other section, but they are still important for you to know. Again, everything here, like the exam, is vendor-neutral. These are all concepts that can be executed in a wide range of data analytic tools and software.

Working with dates

Working as a data analyst, you will quickly learn that date variables are terrible. Every program handles them slightly differently and getting exactly what you want out of them is never easy. There are a few things that will help a little.

Know how to break up a date variable. This changes from program to program, but it is important to learn the basics of how to extract a day, month, or a year from a date variable. You can try to parse it using a simple delimiter, telling the program to break up this variable every time there is a "/" symbol or whatever symbol is used to divide the different parts of the date. It creates a new variable every time the selected delimiter is found, as seen in *Figure 5.18*. Alternatively, you can try to break the variable up after a specific number of characters, but this ends up being a little tricky.

9/17/2022		9	17	2022
9/18/2022		9	18	2022
9/19/2022	⇨	9	19	2022
9/20/2022		9	20	2022
9/21/2022		9	21	2022

Figure 5.18 – Dates and delimiters

Another important thing to know is that you can, and will probably have to, do math with dates. Sometimes you can add or subtract dates from one another, or sometimes there are special functions that will tell you the difference between two dates. No matter what program or language you are using, make sure you learn how to create a derived variable that will tell you the number of days between two dates. It comes up more often than you would expect.

The final tip for working with dates is to figure out a function or feature that will automatically give you the current date or a time date stamp. This may be a simple thing, but it is one of the most important things for you to learn if you will be working with dates.

Conditional operators

We have mentioned conditional logic serval times throughout this book. Not only is conditional logic a staple of programming but it is also featured in every data analytics software to some extent. The concept is simple: you are telling the computer "If this condition is met, then do this." The idea is straightforward, but later it gets more and more complicated as you add multiple conditions or start nesting conditional logic inside conditional logic. In any case, there are a few concepts that are, more or less, standardized across the field. No matter what program or software you choose, these will still be applicable. These are called **conditional operators**, or little bits of code that allow you to create conditional logic. How they are strung together may change, and one program may have more or fewer logical operators, but these four will always be present:

- `IF`
- `AND`
- `OR`
- `NOT`

`IF` starts conditional logic and it specifies the condition that must be met. `AND` means that there are multiple conditions and all of them must be true. `OR` means that there are multiple conditions, but only one of them needs to be true. `NOT` makes the condition a negative and flips the meaning.

Transposing data

Transposing data may be required, depending on what you are doing and the format of the dataset when you receive it. Transposition is simply the act of changing the axis of your data. Columns are now rows and rows are now columns. Let's see what this looks like.

Month	UnitsSold	Color
August	432	Red
August	365	Blue
August	154	Yellow
September	398	Red
September	386	Blue
September	108	Yellow

Month	August	August	August	September	September	September
UnitsSold	432	365	154	398	386	108
Color	Red	Blue	Yellow	Red	Blue	Yellow

Figure 5.19 – Transposing tables

In *Figure 5.19*, we see that what once were columns are now rows and vice versa. While it is a simple concept, things such as this that impact the shape of the data can take some time to wrap your head around in practice. It can get pretty abstract. If you already have this mastered, you can look up information on stacking and melting.

System functions

System functions do not directly impact the data but can be very useful during the data-wrangling process. While there are many and they serve a wide variety of uses, the most common system functions tell you about file paths and your local environment. These may sound like trivial things, but when you are reading data from one location and writing it to another, being able to double-check file paths is very useful.

Summary

In this chapter, we learned about data wrangling and manipulation methods, or how to change your data into something you can use. We covered the different methods of merging data such as joins, blends, concatenation, and appending, and discussed the creation of new variables to help you, such as derived or reduced variables. Next, we discussed parsing data to break it down into little chunks. Then, we covered recoding variables, or changing current variables into a form you can use. Finally, we went over a list of common tools you can find in any analytical program that will help you in your work as a data analyst.

This wraps up *Part 1*, which was all about the preparation of data. Coming up in *Part 2*, we will start looking at analyzing data, and what you can do to learn more from it.

Pretty exciting, right? I'll see you in the next chapter.

Practice questions

Let's try to practice the material in this chapter with a few example questions.

Questions

1. The following picture represents what kind of join?

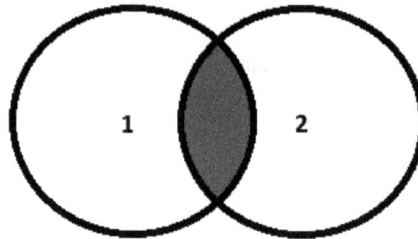

    A.   Inner join

    B.   Outer join

    C.   Left join

    D.   Right join

2. Only using the Distinct Count of a dataset is an example of what?

    A.   Normality

    B.   Recoding

    C.   Transposition

    D.   Reduction

3. The following is an example of what concept?

```
Data = "This is a sentence?"
Data = ["This", "is", "a", "sentence", "?"]
```

A. Transposing

B. Parsing

C. Derived variables

D. Reduction

4. The following is an example of what concept?

Month	UnitsSold	Color	Red	Blue	Yellow
August	432	Red	1	0	0
August	365	Blue	0	1	0
August	154	Yellow	0	0	1
September	398	Red	1	0	0
September	386	Blue	0	1	0
September	108	Yellow	0	0	1

A. Transformation

B. Dummy coding

C. Blending

D. System functions

5. Which of the following is a logical operator?

A. IF

B. NOT

C. OR

D. All of the above are logical operators

Answers

Now we will briefly go over the answers to the questions. If you got one wrong, make sure to review the topic in this chapter before continuing:

1. The answer is: Inner join

   This represents an inner join because only the data that both datasets have in common are added to the new table.

2. The answer is: Reduction

    Distinct Count is an aggregate used to summarize data. Using only this reduces the amount of data you have to process, earning the name reduction variable.

3. The answer is: Parsing

    Parsing is the concept of breaking down large chunks of data into smaller pieces that can be processed and analyzed.

4. The answer is: Dummy coding

    Dummy coding is a specific type of recoding that creates a new variable for every possible outcome of a categorical variable.

5. The answer is: All of the above are logical operators

    The most common logical operators are IF, AND, OR, and NOT.

Part 2: Analyzing Data

This part provides information on some of the most common forms of data analysis, as recognized by the CompTIA Data+ certification exam, and how to perform them.

This part covers the following chapters:

- *Chapter 6, Types of Analytics*
- *Chapter 7, Measures of Central Tendency and Dispersion*
- *Chapter 8, Common Techniques in Descriptive Statistics*
- *Chapter 9, Hypothesis Testing*
- *Chapter 10, Introduction to Inferential Analysis*

6
Types of Analytics

If you look at the objective of the exam, you will find that you need to understand four general categories of analyses: **exploratory data analysis (EDA)**, **performance analysis**, **trend analysis**, and **link analysis**. Here, we will cover everything you need to understand about the types of analyses for the exam. You will learn what EDA is, why it's used, and some common types. For performance analysis, we will explain what it is before jumping into some subcategories. Trend and link analysis are pretty straightforward to understand what they are, and we will discuss some common techniques to execute them. Finally, we will talk about how to choose which category of analysis you will need at any given time.

In this chapter, we're going to cover the following main topics:

- Exploring your data
- Checking on performance
- Discovering trends
- Finding links
- Choosing an analysis

Technical requirements

There is nothing that you absolutely have to do to prepare for this chapter, but there is an example in this chapter using Python in Jupyter Notebook. If you would like to follow along, you can find the `EDA_Example_Data.csv` dataset by following the link provided:

`https://github.com/PacktPublishing/CompTIA-Data-DAO-001-Certification-Guide`

This GitHub link will take you to a repository that contains the data as well as my example code.

Exploring your data

EDA is a general term for a broad category of analyses that are used to understand your data better. Is that definition a little vague? Yes, but the term means slightly different things to different people. Even what analyses fall into this category are up for debate. What is EDA, then? It's dipping your toe into the water to check the temperature before jumping into the data lake.

> **Important note**
>
> It should be noted that some analysts include the cleaning and wrangling processes in what they consider EDA because they are all things you do to prepare your data for use. However, for the purpose of this exam, they are considered separate.

When you first receive a new dataset, before you know what questions to ask or analyses to run, you need to understand some basic information about the data. This can take the form of basic descriptive statistics, simple charts or visualizations, or even simple modeling or machine learning algorithms. What analyses you run may depend entirely on what you need to do with the data. EDA encompasses any preliminary information gathering that you must do before you can jump into what you actually want to know. Maybe you need to know about frequencies, averages, trends, or the relationships between your variables before you can get started.

> **Important note**
>
> As you go through this chapter, you may find that there is some overlap between EDA and the other types of analyses. This is because if your goal is to use the more advanced forms of any type of analysis, you may perform one of the simpler forms as part of your EDA. For example, Structural Equation Modeling is an advanced form of link analysis and isn't any fun to run, so you might do scatter plots as part of your EDA to see whether it is even worth running the advanced model.

Lots of people argue that EDA is only this or only that, but the truth of the matter is that EDA is whatever you need it to be to lay the groundwork for your later analyses.

Common types of EDA

Okay, anything can be an EDA, but a few types are definitely more common than others. Let's start with the most basic and work our way up.

Descriptive statistics

Often, the first thing you need is a suite of simple descriptive statistics. Aptly named, they are there specifically to describe your data. That's all they do. They don't predict or indicate anything. They just give you basic information. There are roughly four categories of descriptive statistics:

- Measures of central tendency
- Measures of dispersion
- Measures of frequency
- Measures of position

Now, these are all more or less what they sound like. Measures of central tendency include things such as **mean**, **median**, and **mode**. We will discuss these in more depth in *Chapter 7, Measures of Central Tendency and Dispersion*. Until then, just know that they help you find the middle of your data. In the context of EDA, these are usually just given as numbers in a table but can be denoted on graphics that show other things.

Measures of dispersion, also to be covered in detail in *Chapter 7, Measures of Central Tendency and Dispersion*, effectively explain how spread out your data is. How much variance you have in your dataset can be an important indicator of how reliable it will be in predictive models. You can expect concepts such as standard deviation, range, variance, min, max, and quartiles. Measures of dispersion are usually visualized as histograms, but box plots are also popular. Again, these values can just be placed in a table.

Measures of frequency will tell you how often specific values come up. This includes concepts such as counts, ratios, or percentages. In EDA, counts are usually just represented in a table but can technically be presented as a bar chart. When it comes to the visualization of percentages, you will sometimes see people use heat maps.

> **Important note**
> All of the charts and graphs listed in this chapter, excluding burndown charts, will be explained in detail in *Chapter 13, Common Visualizations*, where we will go over each visualization in more depth.

Measures of position aren't usually used in EDA. They are all about figuring out how a single value compares to a distribution. These include percentiles, z-scores, and – some would argue – quartiles.

Overall, all of these measures are invaluable in EDA and should be one of the first things you use. It may sound like a lot, but the fact that this information is important has been noted by the people who design data analytics tools. As such, most tools will have some shortcuts to help you get this information faster. For example, if you are using the pandas package in Python 3, you can just use the `.describe()` function, which will automatically give you a table that has a lot of measures done for you.

Relationships

There are several ways to see whether there is a relationship between two variables. These methods range from vague and simple to incredibly specific and complicated. At a high level, the relationships between variables are called **correlations**. When variable A changes, does variable B change as well? We will go over correlation specifically in *Chapter 10, Introduction to Inferential Statistics*. However, for now, know that it is common practice to use a simple visualization such as a scatter plot during EDA. Let's look at one in *Figure 6.1*:

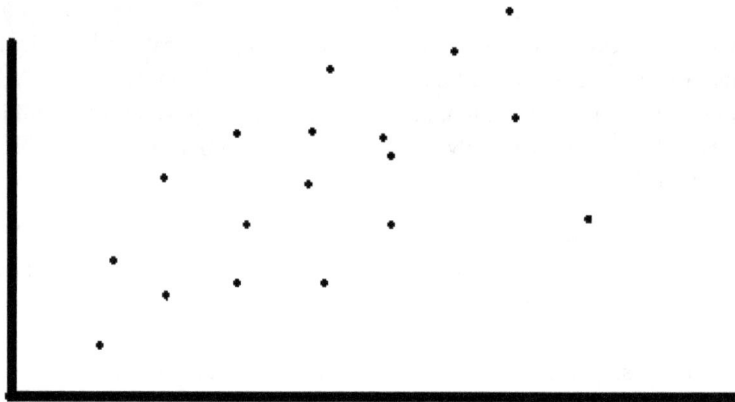

Figure 6.1 – Scatter plot

In *Figure 6.1*, we see an example that gives you a rough idea about whether variables might be correlated so you know whether it is worth taking the time to perform a more in-depth analysis.

Dimension reduction

Dimension reduction, sometimes called feature reduction, like the reduction we have discussed before, is the idea that you should simplify the data as much as you can before trying to use it. The idea is simple, but when people talk about dimension reduction, they are talking about a specific set of advanced multivariate statistics, such as **principal component analysis (PCA)** or **non-negative matrix factorization (NMF)**. These are by no means simple to execute or common at this level of data analytics. Another approach is to use machine learning clustering methods, such as k-nearest neighbors or k-means clustering, to create groups to simplify the data. Let's see how this might work in *Figure 6.2*:

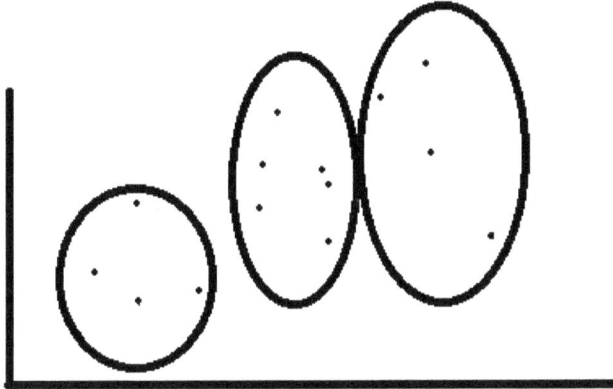

Figure 6.2 – Dimension reduction with clusters

Figure 6.2 is an example of how a clustering method might break data down into groups, then use those groups to generalize instead of using every data point. Nothing about this will be on the exam. I only mention it here because there are those who argue, mostly statisticians, that it is the most important part of EDA.

EDA example

This is all sounding pretty vague, so how about we walk through an example together? We will use a sample of customer data. For the purpose of this example, I will be using Python 3 through Jupyter Notebook.

> **Important note**
>
> Again, you do not need to know how to use any particular software or programming language for the exam. This example is in Python because it is a popular tool and one I personally use. You are encouraged to practice this on your own, but if you would rather, you can simply read along to get a general idea of the process.

If you would like to practice this on your own, I would encourage you to go to the website www. anaconda.com. Anaconda is a free collection of data science tools and modules. Not only will you get Python and even Jupyter Notebook, but it will automatically install the majority of packages that you will need for data analytics. Let's import a package and pull in our data:

```
import pandas as pd
MyData = pd.read_csv("EDA_Example_Data.csv")
MyData.head()
```

The preceding code is pretty straightforward. The first line imports the pandas package and saves it as pd. The second line uses a function from pd to load our data into the program and save it as MyData. The third line is a function that will just display the first few lines of our data. Overall, this imports the package we need, and our data, and gives us a sneak peek of it. The results will look something like this:

	Client_ID	Age_Bracket	AB_Recode	Orders	Total_Spent
0	4682531	41 - 50	4	2	28.0
1	6464235	21 - 30	2	1	23.0
2	9390103	21 - 30	2	10	210.0
3	9346815	< 21	1	4	76.0
4	5672895	21 - 30	2	2	18.0

Figure 6.3 – First five rows of the dataset

The dataset displayed in *Figure 6.3* is a sample of 100 customers from an e-commerce website, though we are only seeing the first five entries. Client_ID is a unique ID number attributed to every client. Age_Bracket shows what age range the client falls into. AB_Recode is a recode of the Age_Bracket variable, and we will talk about why it is there in a moment. Orders is the number of orders placed by the client. Total_Spent is how much money the client has spent in total.

Okay, we have brought in our data and glanced at it. Before we go too in-depth, we should start with a broad view of the data as a whole by executing the following line of code:

```
MyData.info()
```

This one little line will give us a rough description of our dataset as a whole. The results should be as follows:

```
<class 'pandas.core.frame.DataFrame'>
RangeIndex: 100 entries, 0 to 99
Data columns (total 5 columns):
 #   Column       Non-Null Count  Dtype
---  ------       --------------  -----
 0   Client_ID    100 non-null    int64
 1   Age_Bracket  100 non-null    object
 2   AB_Recode    100 non-null    int64
 3   Orders       100 non-null    int64
 4   Total_Spent  100 non-null    float64
dtypes: float64(1), int64(3), object(1)
memory usage: 4.0+ KB
```

Figure 6.4 – Summary of the dataset with info()

In *Figure 6.4*, we see that our data is formatted as a pandas DataFrame, which is fine for what we are doing. We also get a list of every variable, the number of values that are not banked for every variable, what data type each variable is, the counts of each data type, and how much memory it is using.

Because the data has already been cleaned, every variable has all 100 values as non-null. We see that `Client_ID`, `AB_Recode`, and `Orders` are integers, `Total_Spent` is a float, and `Age_Bracket` is an object, which, in this case, means it is a `string` variable.

> **Important note**
> In the majority of programming languages, you will find common variable types. To put it simply, integers are whole numbers, floats are decimals, and strings are text.

It should be noted that if `Total_Spent` was formatted as currency, that dollar sign in front would mean that when we brought it into Python, it would be treated as an object as well, and we would not be able to use it in calculations. There are ways to convert it to a float within Python, but that is something that should be completed before this process, if possible.

For this example, we know that in the future, we will want to perform calculations on `Total_Spent`, so let's look at it a little closer:

```
MyData["Total_Spent"].describe()
```

This line of code performs the `describe()` function on the `Total_Spent` variable found in `MyData`. The results are as follows:

```
count     100.00000
mean      101.52000
std        89.76473
min         2.00000
25%        28.00000
50%        82.50000
75%       137.50000
max       364.00000
Name: Total_Spent, dtype: float64
```

Figure 6.5 – Summary of Total_Spent with describe()

In *Figure 6.5*, we see where the descriptive statistics come in. Here, we have the number of values, the average of the values, the standard deviation, the smallest and largest values, as well as the quartiles. This gives us a lot of information about this variable. If you want, you can repeat this process for `Orders` simply by rerunning the previous line of code, but you will have to replace `Total_Spent` with `Orders`. You could, theoretically, also run it on `Client_ID`, `Age_Bracket`, or `AB_Recode`, but it would be meaningless. `Client_ID` is a key variable and not something you will generally use for analytics, `Age_Bracket` is a categorical variable, and you will only get general counts and what value pops up the most, and `AB_Recode` is a categorical variable pretending to be an integer, so the results won't make any sense.

Okay, well, if the `describe()` function won't help with a categorical variable, what should you use?

```
MyData["Age_Bracket"].value_counts()
```

This is a quick way to take a glance at a categorical variable. It is also a quick way to check for typos in a variable. The results are as follows in *Figure 6.6*:

```
31 - 40    34
41 - 50    21
21 - 30    19
> 50       15
< 21       11
Name: Age_Bracket, dtype: int64
```

Figure 6.6 – Value count results

In the preceding figure, we see every value that is found in `Age_Bracket`, and the number of times each occurs. The values are, by default, sorted from the most frequent to the least. Here, we see that the majority of clients for this website are between the ages of 31 and 40.

Next, we are going to combine a few steps all in one very convenient line of code:

```
EDA_Plots = pd.plotting.scatter_matrix(MyData)
```

This creates a scatter plot matrix of all numerical variables. Also, in the squares on the grid where a variable crosses itself, it displays a histogram that gives you a rough idea of the distribution of that variable. The results are as follows:

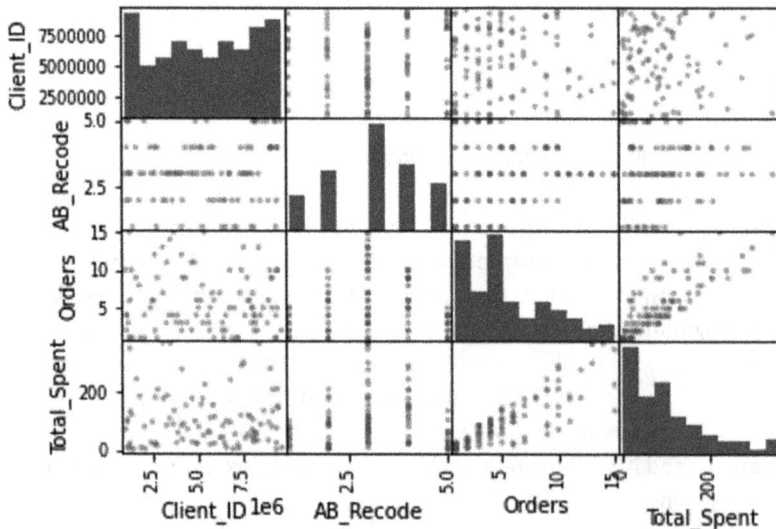

Figure 6.7 – Scatter plot matrix

There is a ton of information packed into this little visualization, as shown in *Figure 6.7*. Let's look at the variables one at a time. The column on the far left is `Client_ID`. Normally, you would not include a key variable in something like this. You would actually create a new dataset, which would be a subset that only included the variables you wanted. However, I left it here to show you what the results would be. The histogram in the top-left corner, which shows the distribution of `Client_ID`, tells us that the distribution is random and does not show any trends, which makes sense because client ID numbers are often generated at random. The next square down, `AB_Recode` / `Client_ID`, shows what it looks like when you include a categorical variable in this matrix, and means nothing. The two squares below that, `Client_ID` / `Orders` and `Client_ID` / `Total_Spent`, show how `Client_ID` relates to `Orders` and `Total_Spent`.

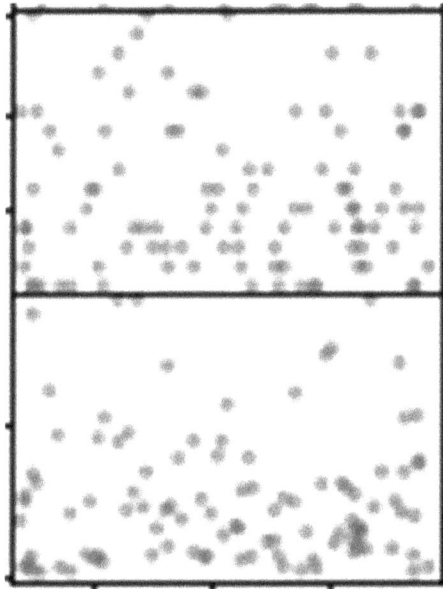

Figure 6.8 – Scatterplots with no relationship

This random scattering shown in *Figure 6.8* does not show anything. There is no relationship between `Client_ID` and anything.

The second column in Figure 6.7 shows `AB_Recode`. Again, you generally do not want a categorical variable in this kind of matrix, but we included it for the histogram, not the scatter plots. This histogram actually shows a pretty normal distribution, as seen in *Figure 6.9*:

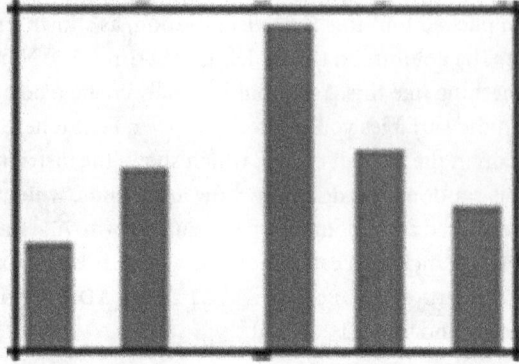

Figure 6.9 – Histogram of AB_Recode

We can do this because there is an inherent order to the values in Age_Bracket; it is ordinal. In other words, we have an order to the categories: <21, 21–30, 31–40, 41–50, and >50. This is because the age brackets are based on the client's age, which is a number.

If you were to try to create a histogram based on a categorical variable such as color, it would be completely meaningless. Blue could be 1, 2, or 27. There is no logic to the order, so there would be no meaning to the shape created by the order.

The third column in Figure 6.7 is Orders. We have already discussed the relationship between Client_ID and AB_Recode, so let's look at the histogram:

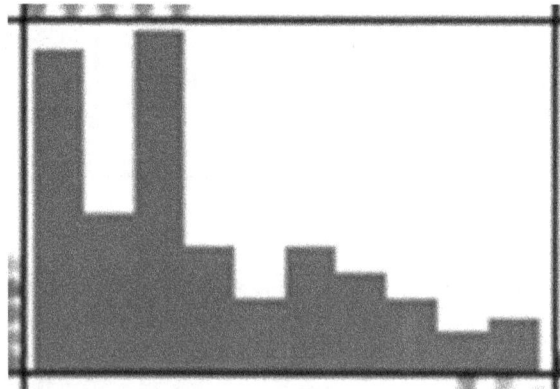

Figure 6.10 – Histogram of Orders

In *Figure 6.10*, we see it is skewed to the right and not a normal distribution, so we should treat it as non-parametric until we try to adjust for the skew. The last box shows how Orders relates to Total_Spent. There is a rough line from the lower left to the upper right, as shown in *Figure 6.11*:

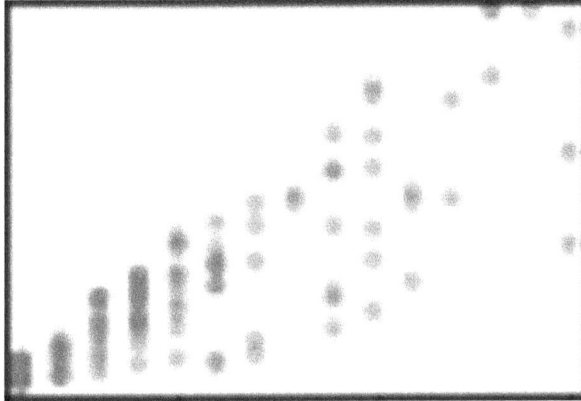

Figure 6.11 – Scatter plot of Orders and Total_Spent

This indicates that there might be a positive relationship between `Orders` and `Total_Spent`. Logically, it makes sense that the more orders a customer makes, the more likely they are to spend more money.

The last column is `Total_Spent` and shows what we would expect at this point. The histogram shows a right skew, it has a positive relationship with `Orders`, a meaningless relationship with `AB_Recode`, and no relationship with `Client_ID`.

This is a lot of information to glean from one line of code. We could go more in-depth if we had specific goals or things to check, but this is a solid amount of basic information from a quick, high-level EDA. We now know enough about our variables to go ahead and get started.

That's what you need to know about EDA. Next, we will move on to performance checks.

Checking on performance

Performance analysis is another slightly open-ended category that is used heavily in the business and industrial fields. At a high level, performance analysis is all about meeting goals. Are you selling as much as you thought you would? Are you producing as much as you expected? Is your team meeting deadlines? Is the manufacturing process as efficient as it should be? Are your competitors' **key performance indicators** (**KPIs**) looking better than yours? The goals that are trying to be met change from field to field, company to company, department to department, and even team to team, so the analyses used to see whether the goals are being met vary drastically. Remember, at a high level, performance analysis is only checking to see whether you are performing as expected. There are a few general subcategories that you might find helpful.

KPIs

We have discussed KPIs previously, but as a quick reminder, they are metrics that are used to gauge performance. Any metric can be a KPI, depending on your circumstances. If your only goal is to make 10 people happy a day, then the number of customers who left with a smile on their face could be a KPI. That said, some KPIs are more common than others. **Return on investment (ROI)** is a classic example because most companies want to know whether they are making money on investments or losing money.

As you can imagine, KPIs are one of the most popular subcategories of performance analytics. There are business analysts who spend most of their time working with these specific metrics. Not only do they have to generate these metrics, but they have to test them to see whether they are reliable and then compare them to the goals. Did they perform better than expected or worse? Is there a statistically significant difference between the goal and the KPI, or is the difference just chance brought on by variance within the sample? What is the percentage difference between the KPI and the goal? These are all things that can be answered by analyzing KPIs.

Project management

Project management has its own set of tools for performance analytics. In this case, the project manager is wondering how this team is performing. Are they completing tasks? Are those tasks on time? Is the productivity consistent throughout the sprint? These are all about keeping a team on track. You could argue that these still use metrics, which could be considered KPIs for the team, but the tools and analyses used are so different and specialized that it is often considered that they are in their own subcategory.

A lot of project, team, or task management software will automatically calculate a lot of these metrics for you. For example, Jira will automatically generate burndown charts, burnup charts, velocity charts, and sprint reports. Let's see what one looks like in *Figure 6.12*:

Figure 6.12 – Burndown chart

In *Figure 6.12*, we see a simplified example of a burndown chart. Even when these are created manually, it is often the job of the project manager to deal with these specific analytics and not the data analyst.

Process analytics

Process analytics is about looking at a process that involves multiple steps. This is used heavily in fields such as manufacturing, where how efficient a process is makes a huge difference to the company as a whole. Not only does it look at the final outcome, but it looks at every step along the way and how it interacts with the next step. Here, you might find analyses such as process capability or process control. The goal of process analytics is focused on the efficiency of a series of steps instead of one large KPI at the end.

That wraps up progress checks. Let's jump right into trend analysis.

Discovering trends

Trend analysis, sometimes called **time series analysis** and **projections**, does exactly what it sounds like, **analyze trends**. Again, this topic is very broad because it is used in so many fields in so many ways. The general idea is that you are looking at a variable over time to see whether there are any patterns or whether you can predict what will happen in the immediate future. The further into the future you go, the less accurate the predictions become.

For example, you know your own weight in pounds. It is a decently safe bet to say that in 1 second, your weight may go up or down a pound because of measurement errors, but it probably won't change. What about in a week? You could probably say your weight won't change, but it could theoretically go up or down 5 pounds. A month? Give or take 15 pounds. A year? Give or take 50 pounds. A decade? Give or take 100 pounds. I think you see where this is going. The further into the future you get, the more chance there is for things to change and the less accurate your predictions can be.

Trend analysis is used heavily in fields such as finance, accounting, or investment, but it can be used in any field. Yes, trying to predict how stocks will change or how much money your company will bring in next year are important things. However, trend analysis also includes things such as sentiment analysis or recommendation engines that can predict how popular things are or whether or not you are likely to purchase an item.

Simple to extremely complicated, trend analyses run the whole gamut. A lot of data analytics tools will have a feature that will automatically add a trendline for you. That is technically a trend analysis. Extending that trendline into the future is called **forecasting**. Let's see what this looks like in *Figure 6.13*:

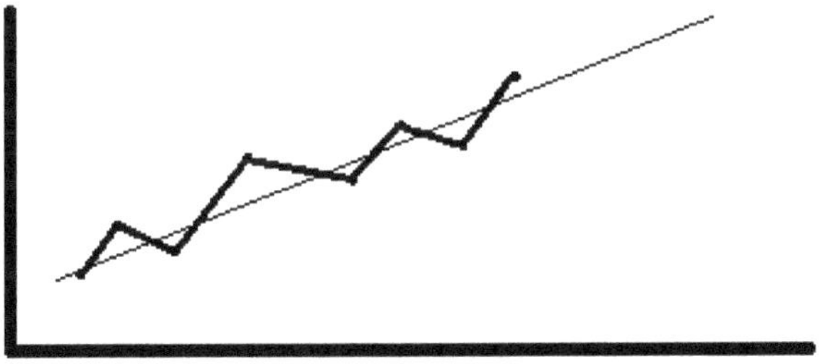

Figure 6.13 – Forecasting

In *Figure 6.13* we see what this might look like. The thicker line is the historical data that has been recorded, while the thin line is a trend line that gives a general idea of where it might be going. There are several different ways to do this, but often this is some type of linear regression, which we will discuss in detail in *Chapter 10, Introduction to Inferential Statistics*.

Finding links

Link analysis on this test is not what you would expect. If you try to look up link analysis on your own, you are going to be very confused. You will probably find something about it being part of network theory and the relationship between nodes, or you may find information on how Google analyzes hyperlinks. You might even find link analysis and criminology or link analysis and customer data analytics. The list goes on and on, and none of these are the link analysis that CompTIA wants you to know. Oops.

For the purpose of this exam, link analytics is a general category of tools and techniques that explain the relationships between variables. These range in terms of difficulty and accuracy. On one end of the spectrum, you have things such as scatter plots that are quick and easy but not particularly accurate. A lot of data analysts still use these as a place to start to see whether more accurate methods are required. In the middle, you have correlation analyses. There are several different kinds of correlation analyses, but they all follow the same basic concepts. They are accurate and within the reach of a data analyst. Correlation of one sort or another makes up a large percentage of what people actually use. At the far end of the spectrum, we have Structural Equation Modeling. This is an advanced multivariate technique that tells you much more than a basic correlation, but it is also a pain to actually use. In the grand scheme of things, these are beyond most data analysts, so you don't need to worry about them.

Choosing the correct analysis

With advances in coding and data analytics programs, it can take longer to choose the correct analysis than it does to actually run it. Now there are data analytics positions where you will never have to choose because it has already been decided for you; run this code on this data every Thursday to generate a report using this template. There is nothing wrong with this, but the majority of data analytics positions will have some element of choice involved.

Now, you may have to choose between specific analyses in the exam, and we will cover those later in *Chapter 10, Introduction to Inferential Statistics*. A lot of the information in this section is just to help you along in your future career as a data analyst.

Why is choosing an analysis difficult?

There are dozens of different analyses for different situations, and many of those have several specialized types, and even among the specialized types, we may have variations or modifications. You can even use some analyses that only exist to modify other analyses. To make matters worse, there are often debates about which is more effective in certain situations. There are even some aspects that may be a matter of preference. If two analyses do the same thing, but one is more conservative and is more likely to reject things that should be accepted, while the other is more daring and is more likely to accept things that should be rejected, which is the right choice? The answer is that it depends.

To make things even more fun, the code and the software will usually run no matter what, which means you will still get an answer even if you are using the wrong analysis or you don't meet the requirements for that analysis. You won't even know that anything is wrong and will blithely report false information. Long story short, choosing the correct analysis without a lot of experience is a pain.

Before we jump into how to deal with this, we need to talk about prerequisites.

Assumptions

You often hear people saying that you should not make assumptions, but the fact of the matter is that they are absolutely necessary, and you make them all the time. There is a huge amount of sensory data flooding your brain every moment, so much that you literally cannot process it all and still function. How, then, are you reading this book? Your brain automatically makes assumptions about what is important, and a lot of the rest is filtered out. Even what is stored in your memory is largely filtered based on what your brain assumes you will want to know later.

Beyond that, you make choices based on assumptions all the time. You go to work because you assume that they will stop paying you if you stop going, which is a safe assumption. You also assume that you will want that money in the future and that you will have the time and ability to spend it. You even assume that the great robotic uprising will not happen before your next paycheck.

While you are making all of these assumptions, analyses are also making assumptions. Every statistical analysis is, at its heart, an equation. Now, if analyses did not make assumptions, you would have to rewrite every equation every time you wanted to use it to account for your specific data. That sounds like a lot of work. Luckily, statistical analyses make assumptions. They assume specific things about your data are true.

If the assumptions are not true, but you run the analysis anyway, the results are not reliable. It is a shot in the dark whether they are accurate or not. It depends on the assumptions they don't meet and blind luck. They could be accurate totally by accident, but the odds are not in your favor. Long story short, make sure you know what assumptions an analysis is making and make sure you meet those assumptions.

Making a list

Before you jump into choosing an analysis, there is something you should do. You will need to create a list of analyses. Every time you learn a new analysis, you will add it to this list. It doesn't matter whether it is a `.txt` file on your computer or a physical notebook. You can even keep this list in your unicorn diary; I won't judge. You just need to write the analyses down somewhere and keep them together. When you do write them down, every entry will need very specific information:

- What is the purpose of the analysis?
- What variables does it require?
- What assumptions does it make?

You will need to answer these three questions for every entry. The purpose of the analysis can be a high-level summary in your own words, just so you know at a glance what it does. What variables it uses can include information on the number of independent or dependent variables or whether it needs a quantitative variable or a qualitative variable. The assumptions are just a list of prerequisites, which we just discussed.

There is nothing stopping you from adding additional information if you like. Maybe you want to add a link to an example of the code to use to actually run it, or maybe you just want to add a doodle of the researcher who came up with it. You can organize it alphabetically or by analysis type – whatever is easiest for you.

Finally choosing the analysis type

Okay, are you ready for the magic bullet that will instantly help you choose the perfect analysis every time?

Too bad. It doesn't exist.

However, there is a process that helps:

1. Know your goal.
2. Know your variables.
3. Check your assumptions.

Let's break this down. First, know your goal. When picking a specific type of analysis, you need to have a clear objective. This is usually to answer a specific business question, and we will talk more about how to write those questions in *Chapter 9, Hypothesis Testing*. Are you comparing two things? Are you seeing whether two variables are related? Are you trying to predict something? Even simple questions such as this can push you in the general direction. Using the analysis types covered in this chapter, you can narrow down your search.

Now it's time to use your statistical analysis list! Go through and find any analyses that have a purpose that is similar to your goal. Sometimes you will only have one, which keeps things simple. Next, we have to check whether that one will work or not.

Step 2 is to know your variables. This means, yes, check the type of analyses in your list to see whether you have the variables those analyses require, but it also means knowing what variables you have available to you. Trying to do forecasting without some time variable will get you nowhere. Sometimes, if you don't have the variables you need, you can find or make them, but sometimes it means there is no way to do that analysis, and you have to find another approach.

Okay, now you have a short list of the various types of statistical analyses that have the correct general purpose and require variables that you actually have. This brings us to step 3, check your assumptions. This step can take the longest, and occasionally you will need to run a type of analysis to see whether you meet the assumptions to run another analysis. That is why this is the last step, so you check assumptions on the smallest number of analyses possible.

This approach to choosing an analysis does require you to go out of your way to learn new analyses and record them in a very specific way, but it is a simple and practical approach to a complicated process. As time goes on and you gain enough experience, you will be able to skip this process more and more. Until then, stick with it and start making that list!

Summary

This chapter was all about analysis types. First, you learned that EDA is a set of quick, preliminary analyses designed to give you some basic information about your data. Using it, you will have a better idea of what questions to ask, how to ask them, and whether it is worth trying more advanced analytics, and it gives you a general idea of what your data is like. You also covered common types of EDA, such as descriptive statistics, relationships, and dimension reduction.

While talking about performance analysis, or analyses designed to check the progress of a company, team, or process, we covered KPIs, which focus on comparing key metrics; project management, which focuses on tracking how a team is performing; and process analysis, which tracks the efficiency of a multi-stage process.

Trend analysis is all about looking at things over time. Often, this is used to forecast, or predict, what a number will do in the future, but it can also be used to track how a group of people feel about something, or even recommend products to customers.

Link analysis is about finding relationships between variables. If one variable changes, does it impact this other variable? Is there a link between a customer's hair color and whether or not they buy eye makeup? These analyses strive to find out.

Finally, this chapter wrapped up by going over how to choose the correct statistical analysis. You learned that choosing the correct one can be difficult. Assumptions are requirements that need to be met for an analysis to be accurate. We discussed making a list of every analysis you know, containing its purpose, its variables, and its assumptions. Finally, we walked through the process of using this list to find the correct analysis by knowing your goals, knowing your variables, and checking your assumptions.

In the next chapter, we will start going over the specific calculations you will need for the exam, so get ready to brush off your arithmetic skills!

Practice questions

Let's try to practice the material in this chapter with a few example questions.

Questions

1. What is the purpose of EDA?

    A. Checking on the efficiency of a program or project
    B. Understanding basic information about your data before performing more advanced analytics
    C. Understanding how or whether different variables relate to one another
    D. Checking on the progress of a variable over time

2. A manager of a small team wants to gather some metrics to track how the team is doing and where they have room for improvement. What type of analysis would they need to do?

    A. Exploratory data analysis
    B. Link analysis
    C. Performance analysis
    D. Trend analysis

3. Forecasting falls into which analysis type?

    A. Link analysis
    B. Performance analysis
    C. Exploratory data analysis
    D. Trend analysis

4. Scatter plots, correlation, and structural equation modeling all fall into which analysis type?

    A. Performance analysis
    B. Trend analysis
    C. Link analysis
    D. Exploratory data analysis

5. In reference to statistical analyses, what are assumptions?

    A. A list of prerequisites that must be met
    B. You should never make assumptions, confirm everything
    C. A list of things to avoid
    D. None of the above

Answers

Now we will briefly go over the answers to the questions. If you got one wrong, make sure to review the topic in this chapter before continuing:

1. The answer is: Understanding basic information about your data before performing more advanced analytics

    Exploratory data analysis is your chance to get a feel for your data and learn the basics about it. It is important to take this step, so you know what you can or should do next.

2. The answer is: Performance analysis

    Performance analysis can include a wide range of different statistical methods, but they generally focus on using key metrics to track progress and performance.

3. The answer is: Trend analysis

    Forecasting is a specific kind of trend analysis that looks at predicting the future of a metric by looking at what it has done in the past.

4. The answer is: Link analysis

    Scatter plots, and even correlation, technically can be used in EDA but they are still considered part of link analysis. Also, structural equation modeling is an advanced technique that would not be used in EDA.

5. The answer is: A list of prerequisites that must be met

    Assumptions are things that the equation assumes about the data to be true, and the data must meet those expectations in order to give an accurate result.

Measures of Central Tendency and Dispersion

This is where you finally get to start analyzing data! For better or worse, there are some calculations that you will have to be able to perform in the actual exam. This is a part that a lot of people get stuck on because these aren't things that data analysts calculate manually, but you do need to know them for the exam. Online testing centers should have a simple calculator built into the calculation questions, and an online document for taking notes. However, you will not be provided with the equations you are expected to know. Time to dust off some of those old math skills!

Before we jump into the equations, we will start by going over what a distribution is and some common distributions you might come across. Next, we will start with the simplest math, talking about measures of central tendency, how to calculate them, and when to use which. Then, we will go over measures of dispersion. We will focus on ranges and quartiles first. Then, we will move on to variance and finish up with standard deviation. There is a lot to cover, so let's get started!

In this chapter, we're going to cover the following main topics:

- Discovering distributions
- Understanding measures of central tendency
- Calculating range and quartiles
- Finding variance and standard deviation

Discovering distributions

Distributions are often discussed when statistics come up; they have even been mentioned earlier in this book. The question is, what are they? A statistics class will tell you something like the following: a distribution is a function that describes a line that depicts the probabilities of any theoretical outcome that occurs, based on the evidence of a study. Is this easily understood and actionable information? Not really.

Effectively, you can think of a distribution as a model of historical data. It tells you how likely a specific value is when compared to everything you have collected before. This has all kinds of uses. You can predict the probability that a new entry will be a specific value, or in a range of values. You can take a value and see how it compares to the rest of your information. It even turns out that the shape of your distribution can tell you all kinds of things. Additionally, several statistical methods require a specific distribution as one of their assumptions.

There are specific methods for finding the exact distribution of your data, but you can get away with making a histogram and eyeballing it most of the time. Is this the ideal approach? Can you tell whether something is perfectly within the allowable range of variables, which will tell you whether it is a specific distribution or not? No, but this is not a book on statistical rigor for research methodology. Data analysts will rarely have to do more than make a histogram.

This is all pretty theoretical, so let's look at an example. You follow a soccer team, the Fighting Goldfish, and write down the number of goals they make for 100 games: 10 they didn't score at all, 20 they scored 1 goal, 50 times they scored 2 goals, 20 times they scored 3 goals, and 10 they scored 4 goals. First, let's try to visualize this through the following diagram:

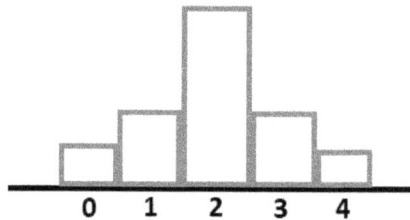

Figure 7.1 – Histogram of Fighting Goldfish goals

In *Figure 7.1* here, we see this data displayed as a histogram. Technically, the distribution is the function describing the line.

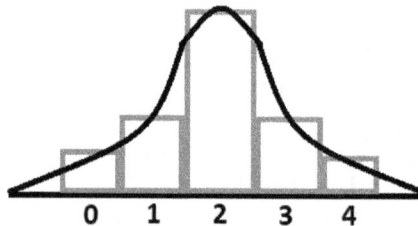

Figure 7.2 – Distribution of Fighting Goldfish goals

In *Figure 7.2*, we see that while the curve of the line describes the distribution, if you point to any specific point on that line, you can find the probability of a given value. However, you will notice that the histogram does a pretty good job of telling us what the shape of the distribution is. Now that we have this data set up, one way or another, we can answer all kinds of questions. What is the probability of the Fighting Goldfish scoring two goals in their next game? 50%. In their last game, the Fighting

Goldfish scored four goals; how does that compare to their other games? That is higher than 70% of their other games. There are all kinds of things that you can learn from a probability distribution, including how much variance there is and how reliable your data is.

Earlier, we mentioned that you can learn some things from the shape of the line. Some shapes naturally come up more often than others and are considered common distributions. Depending on what sort of data you are using, some distributions may be more common than others. Let's go over a few that you might run into.

Normal distribution

Gaussian – does it ring a bell? If you didn't get that joke, trust me, it's hilarious. A Gaussian distribution, often called a **normal distribution**, looks like a bell and is often described as a **bell curve**. In *Figure 7.3* here, we get a rough idea of the shape:

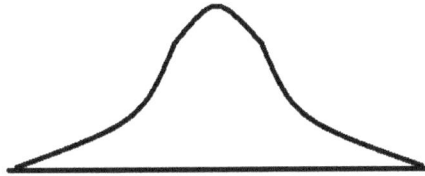

Figure 7.3 – Normal distribution

Many statistical analyses require this distribution because it is the most common and pops up all the time. If you're curious, the formula for the function can take many forms, but is often roughly something like this:

$$g(x) = \frac{1}{\sigma\sqrt{2\pi}} \exp\left(-\frac{1}{2}\frac{(x - \mu)^2}{\sigma^2}\right)$$

The basic idea behind this distribution is that, ideally, the mean, median, and mode are all in the center and that is the highest point in the curve. As you move away from the mean, left or right, the probability decreases evenly on both sides, so both sides are equal. This is what gives it the characteristic bell shape.

Uniform distribution

A uniform distribution is very boring. It is, quite literally, a flat line. If you were to create a histogram of a dataset of 1, 2, and 3, you would get a uniform distribution.

Figure 7.4 – Uniform distribution

In *Figure 7.4* here, we can see what a uniform distribution looks like. What does this tell us? That every value has the exact same chance of happening. There is a maximum value and a minimum value, but everything between has happened roughly the same number of times in the past. When you have a uniform distribution, there is not a lot to do besides watch to make sure it doesn't break. If you suddenly get a value above your maximum or below your minimum, or if you suddenly start getting the value in the middle more than the others, it means something has drastically changed your data and it will probably impact your other variables as well.

Poisson distribution

Poisson is a pretty specialized distribution. This is the number of times an event happens over a fixed, repeating period. It's usually time but does not have to be. When I originally taught this, the example used was counting every occurrence of roadkill for every mile down a certain road. With this, you can give the probability of the number of occurrences of roadkill down the next mile of the road. Neat, huh?

Perhaps a more relevant example: you are looking at how many hits a website gets every hour, or how many products are sold a day. These are all based on a count over a repeating interval, usually time.

That said, the shape of a Poisson distribution depends on Lambda (λ), which is defined by the rate, which is, in turn, defined by the average. Long story short, this one can have different shapes:

Figure 7.5 – Possible Poisson distributions

In *Figure 7.5* here, we can see that this is a tricky one, and it is defined more by what the variables are than by the shape.

Exponential distribution

Exponential distributions are, as the name suggests, distributions that increase or decrease exponentially. These are often described as being related to Poisson distribution because they are used to describe the waiting time between events. However, they do exist on their own. If you see any variable that changes with a sharp curve, it might be exponential:

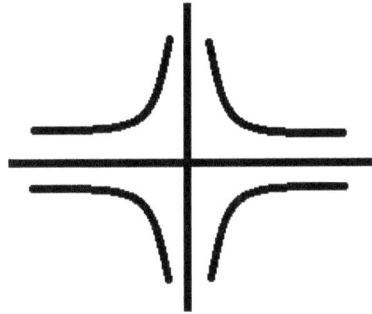

Figure 7.6 – Exponential distributions

In *Figure 7.6* here, we can see that this does have multiple shapes, but they are all the same basic shape with different orientations.

Bernoulli distribution

Bernoulli distributions are another specialized distribution, but they are very simple and easy to identify. Just remember that Bernoulli distributions are Booleans. There are only two possible outcomes: true or false. You can call these outcomes whatever you want: pass or fail, success or failure, 1 or 0, yes or no, and so on. It doesn't matter what you use as long as it is a binary variable and all values are either one thing or the other. This is used, quite simply, to give the probability of one of the outcomes. This is also only run once.

Let's say you are a fan of model rockets. For those of you who don't know, a model rocket is a cardboard tube with a plastic tip that you stuff with explosives to send flying in the air. If stuffed improperly, there is a chance that the rocket will just sit there, catch fire, explode, or some combination of any of these things. Any outcome that does not involve the rocket safely being launched into the air in one piece is considered a failed launch. Hypothetically, for someone without experience, there is a 50% chance that your rocket will successfully launch and a 50% chance that it will explosively fail on any one launch!

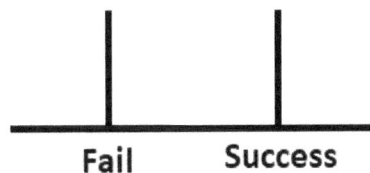

Figure 7.7 – Bernoulli distribution

In *Figure 7.7* here, we get a general idea of what this looks like as a distribution. It is probably safest to not stand too close to the rocket when you try to launch it.

Binomial distribution

A binomial distribution is like a Bernoulli distribution, but the experiment is repeated. Instead of launching one rocket, you launch ten! This changes things a little. For example, instead of looking at success and failure, you are now only looking at the probability of success. To be more accurate, you are looking at the probability of a specific number of successes.

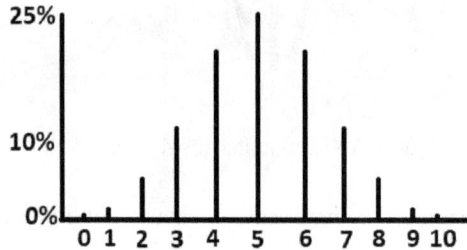

Figure 7.8 – Binomial distribution of rocket launches

In *Figure 7.8* here, we can see our binomial distribution. If we launch 10 rockets and each one has a 50% chance of success, what is the probability that we will have exactly 5 successful launches? Roughly 25%.

Note that binomial distributions do not have a set shape. If we look at *Figure 7.7*, we can see that this is normally distributed, but that is only because we have a success rate of exactly 50%. The shape of this distribution changes based on the probability that any one run is successful.

Skew and kurtosis

Skew is a measure of crookedness. If a distribution leans to the left or to the right, it is skewed. If it leans too far, then the distribution can no longer be considered normal. This makes sense since one of the defining characteristics of a normal distribution is that the sides are equal, so if one side is longer than the other, it no longer meets this criterion.

Figure 7.9 – Negative or left skew

In *Figure 7.9* here, we can see a distribution that is skewed to the left. This is also called a negative skew. You can remember the name because the long tail points toward the skew. The tail is pointing to the left or toward the negative numbers, so it is a left or negative skew:

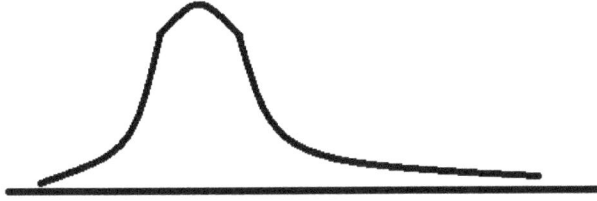

Figure 7.10 – Positive or right skew

In *Figure 7.10* here, we can see, you guessed it, a distribution that is right-skewed, or positively skewed. Again, the long tail points to the right or toward the positive numbers. Lots of things can cause a skew, including, but not limited to, outliers that pull the probability to one side or the other.

If skew describes your distribution being crooked to the left or right, kurtosis is it being crooked up or down. To be more precise, it is looking at the difference in the tails from a normal distribution:

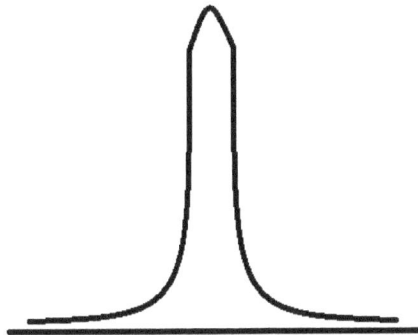

Figure 7.11 – Leptokurtic distribution

In *Figure 7.11* here, we can see a leptokurtic distribution. That means it is tall and skinny, with tails smaller than a normal distribution. It also means that all of the values are closely packed at the center line. You can remember this name because the distribution "lept" into the air! Like "leaped?" It's a pun.

Figure 7.12 – Platykurtic distribution

In *Figure 7.12* here, we see a platykurtic distribution. This is a short, wide distribution, with larger tails than a normal distribution. In this case, the data is much more spread out and is only slightly more likely to be near the center line. You can remember this name because it is short and flat like a plate. As a side note, a normal distribution is said to be mesokurtic, which you can remember because meso literally means middle.

You now have a rough understanding of skew and kurtosis. Does the distribution lean left or right? Is it tall or squat? That said, the ideal data may not actually exist in the wild. Your data is allowed to lean a little bit, or be a little tall, and still be considered normal. There are actually tests that will give you specific numbers for skew and kurtosis and an acceptable range for what is considered normal, but at the end of the day, it is a judgment call. Does it look like a bell curve or is it so deformed in any direction that it is no longer normal? That's up to you.

To be perfectly clear, you should be familiar with the concepts of skew and kurtosis for the exam, but you will not be asked whether a specific skew value is too high or how to normalize a distribution that is skewed.

Understanding measures of central tendency

A measure of central tendency is a summary of a dataset in a single number. The name comes from the fact that these values should be in the center of your distribution and should be the value with the highest probability of occurring. Overall, because they are useful summaries, they are often used as variables in other statistical analyses. You can use them by themselves. It is easiest to compare two averages than to compare every value in two datasets.

There are a few different ones, but in common practice, you will only come across three:

- Mean
- Median
- Mode

You may have heard of these before, but let's go over them briefly.

Mean

The mean is also called the average. To find the average, you take the sum of the values and divide them by the number of values. This may be simple, but let's break it into steps:

1. Find the sum of the values.
2. Divide the sum by the number of values.

Let's try a quick example. You have three puppies. The puppies weigh 22 lb, 26 lb, and 24 lb:

1. Find the sum of the values.

    For our example, we will add together 22, 26, and 24:

    22 + 26 + 24 = 72

    So, our sum is 72.

2. Divide the sum by the number of values.

    We have three puppies, so we have three values:

    72 / 3 = 24

    The mean weight of our puppies is 24 lb.

    You can practice this process on your own with the following dataset: 12, 7, 11, 6, 9.

Median

The median is the middle of your data. Aren't they all? Well, yes, but the median is literally, if you put all of your values in ascending order, the value that falls in the exact middle. The steps are as follows:

1. Arrange values in ascending or descending order.

2. Find the value that is in the exact middle.

3. If your dataset has an even number of values, find the mean of the two in the middle.

Let's go back to our puppies. As you may recall, their weights (lb) are 22, 26, and 24:

1. Arrange values in ascending or descending order.

    If we put them in ascending order, we get 22, 24, 26.

2. Find the value in the exact middle:

    22, **24**, 26

    24 is in the middle.

Because we had an odd number of values, that is the end and 24 is our median. Let's try this process again with the addition of a fourth puppy who weighs 25 lb, so our new dataset is 22, 26, 24, 25:

1. Arrange values in ascending or descending order.

    If we put them in ascending order, we get 22, 24, 25, 26.

2. Find the value in the exact middle:

    22, **24, 25**, 26

    Now, 24 and 25 are both in the middle, so we continue to the next step.

3. If your dataset has an even number of values, find the mean of the two in the middle.

    So, we have to find the average of 24 and 25:

    24 + 25 = 49

    49 / 2 = 24.5

    The median for this dataset is 24.5 lb.

You can practice this process on your own with the following dataset: 6, 12, 8, 72, 1, 15.

Mode

Mode is all about frequency. It is literally whatever number occurs most often in your dataset. This one is a little different from the others. You can have no mode if no value is repeated, or you can have multiple modes if two values happen the same number of times. The steps are as follows:

1. Arrange values in ascending or descending order.
2. Count the occurrences of all repeating values.
3. Compare the number of occurrences for every repeated value.

We will use the puppy data again, because everyone loves puppies. However, we will add a few more puppies! Our puppy data is now 22, 26, 24, 25, 25, 24, 23, 25, 22. Whew! That is a lot of puppies:

1. Arrange values in ascending or descending order:

    22, 22, 23, 24, 24, 25, 25, 25, 26

2. Count the occurrences of all repeating values:

22	23	24	25	26
2	1	2	3	1

3. Compare the number of occurrences for every repeating value.

    25 has the highest number of occurrences.

    Our mode is 25 lb.

    Hypothetically, if we had 1 fewer 25, then 22, 24, and 25 would all be modes because they would all be tied for the highest frequency.

You can practice this process on your own with the following dataset: 7, 3, 2, 4, 3, 4, 5.

When to use which

Okay, so what is the point of having different measures? Do you have to find all of them every time? The answer is no. Some of them work better with certain kinds of data. Let's look at each in turn.

When to use the mean

Mean is the default; when in doubt, it is a safe bet. It works with normal distributions perfectly. It can work with other distributions but is easily pulled in the wrong direction by outliers, so skewed data or asymmetrical data can give you incorrect values.

When to use the median

The median is an underrated metric, and some people use it for everything because it is immune to outliers. It doesn't matter whether there is a value at the end that is 1,000 times higher than any other value; it still only counts as 1 value above. This makes it ideal for working with skewed or asymmetrical data, where a mean can't. Medians are also great for working with ordinal variables. If you recall, an ordinal variable is a scale that uses regular intervals, so medians naturally work well with finding the middle.

When to use the mode

The mode is not very popular, but it can do something that others can't. It can work with nominal variables. Recall that nominal variables are categorical variables that have no inherent order. You could try to take the average or the median of colors, but it would be a very silly thing to do. However, since the mode works with frequencies, it is ideal for handling nominal variables.

Calculating ranges and quartiles

Range and quartiles, as well as variance and standard deviation, are considered measures of dispersion. As the name suggests, these are all ways to find out how dispersed or spread out your data is. Is your data random and widely spread, or are the points tightly clustered around the mean? Not only are these values used in more advanced calculations but they are also very useful in and of themselves. If you recall, several of these were common in EDA because it is basic information that has many uses.

Ranges

A range is the simplest measure of dispersion. The idea is simply to know how far spread your dataset is. This is designed to be quick and easy, but not the most useful measure. The steps are as follows:

1. Arrange values in ascending or descending order.
2. Identify the minimum and maximum values.
3. Subtract the minimum from the maximum.

Let's look at an example. You are hired by a small chicken farm and you are tracking the number of eggs produced every day for 10 days. The dataset is as follows: 12, 18, 10, 22, 15, 25, 16, 17, 14, and 19:

1. Arrange values in ascending or descending order:

   10, 12, 14, 15, 16, 17, 18, 19, 22, 25

2. Identify the minimum and maximum values.

   Luckily, now that we have ordered them, these are the values on the far left and the far right:

   10, 12, 14, 15, 16, 17, 18, 19, 22, **25**

3.   Subtract the minimum from the maximum:

$$25 - 10 = 15$$

Just like that, we have a range of 15.

You can practice this process on your own with the following dataset: 4, 6, 5, 2, 6, 8, 4, 6, 1, 4.

Quartiles

Quartiles divide your data into quarters. To divide your data up into four pieces means that you have three points of division as seen in *Figure 7.13* here:

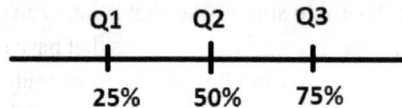

Figure 7.13 – Quartiles

These points of division are called the lower quartile (**Q1**), the middle quartile (**Q2**), and the upper quartile (**Q3**). Normally, when this is taught, you learn a slightly different equation for each quartile that will give you the position in an ordered dataset of each quartile. Then, you can look up each value by its position. This isn't a bad method, but there is an easier way that I will teach you:

1.   Arrange values in ascending or descending order.
2.   Find the median – Q2.
3.   Split the dataset at the median.
4.   Find the median of the lower dataset – Q1.
5.   Find the median of the upper dataset – Q3.

You already know how to find the median, so let's jump right in with our egg data – or is it chicken data? Which came first? The dataset is as follows: 12, 18, 10, 22, 15, 25, 16, 17, 14, and 19:

1.   Arrange values in ascending or descending order:

10, 12, 14, 15, 16, 17, 18, 19, 22, 25

2.   Find the median:

$$\frac{(16 + 17)}{2} = 16.5$$

The median is 16.5, so Q2 is 16.5.

3. Split the dataset at the median:

    Lower dataset: 10, 12, 14, 15, 16

    Upper dataset: 17, 18, 19, 22, 25

4. Find the median of the lower dataset:

    Lower dataset: 10, 12, **14**, 15, 16

    Q1 is 14.

5. Find the median of the upper dataset:

    Upper dataset: 17, 18, **19**, 22, 25

    Q3 is 19.

Therefore, our lower quartile is 14, our middle quartile is 16.5, and our upper quartile is 19.

You can practice this process on your own with the following dataset: 3, 8, 4, 9, 2, 6, 5, 9, 4, 7.

Interquartile range

An interquartile range is a measure of dispersion that combines the two techniques you just learned. You are literally just finding the range of the quartiles. This should be a quick one:

1. Find the quartiles.

2. Subtract Q1 from Q3.

    Since we just found the quartiles of our dataset in the last example, this should be easy:

1. Find the quartiles:

    Q1 = 14, Q2 = 16.5, Q3 = 19

2. Subtract Q1 from Q3:

    19 – 14 = 5

    The interquartile range is 5.

You can practice this process on your own with the following dataset: 6, 4, 2, 5, 6, 3, 4, 1, 4, 6.

Finding variance and standard deviation

Variance and standard deviation are very popular. They are a little bit more complicated to perform by hand, not that you would ever perform them by hand if you didn't have to for the exam, but they are a much better measure of how dispersed your data is. Instead of giving you a rough idea based on the range, these tell you the average distance of every point from your mean.

Variance

Variance is a measure of dispersion that looks at the squared deviation of a random variable from the mean of that variable. This equation looks a little scary, but we will break it down step by step:

$$s^2 = \frac{\Sigma(x^i - \bar{x})^2}{n - 1}$$

It should be noted that this denominator (n-1) is used for samples. If you are using an entire population, then the denominator is just (n). Let's go over this briefly. s^2 is the sample variance, x^i represents the value of each observation, $\bar{x}$ represents the mean of the sample, and n is the number of data points in your dataset. Let's go ahead and go through the steps:

1. Find the mean of the dataset.
2. Subtract the mean from each data point.
3. Square the results from the previous step.
4. Find the sum of the previous step.
5. Divide that sum by the number of data points minus 1.

Some of these steps may not sound clear, so let's jump right into the example. Now, you work for a website that sells specialized boats. The values represent how many boats are sold in a week: 3, 6, 4, 7, 5, 1, 9, 5, 4, and 6:

1. Find the mean of the dataset:

$$\frac{(3 + 6 + 4 + 7 + 5 + 1 + 9 + 5 + 4 + 6)}{10} = 5$$

   The mean is 5.

2. Subtract the mean from each data point:

   (3 – 5), (6 – 5), (4 – 5), (7 – 5), (5 – 5), (1 – 5), (9 – 5), (5 – 5), (4 – 5), (6 – 5)

   becomes

   (-2), (-1), (1), (2), (0), (-4), (4), (0), (-1), (1)

3. Square the results from the previous step:

   $-2^2, -1^2, 1^2, 2^2, 0^2, -4^2, 4^2, 0^2, -1^2, 1^2$

   becomes

   4, 1, 1, 4, 0, 16, 16, 0, 1, 1

4. Find the sum of the previous step:

   $4 + 1 + 1 + 4 + 0 + 16 + 16 + 0 + 1 + 1 = 44$

   The sum is 44.

5. Divide that sum by the number of data points minus 1:

$$\frac{44}{(10 - 1)} = 4.9$$

The variance for this dataset is 4.9.

You can practice this process on your own with the following dataset: 10, 8, 12, 11, 9, 9, 8, 12, 11, 10.

Standard deviation

You may have noticed that variance is represented by s^2. You will be happy to know that standard deviation is just s. That's right – standard deviation is a measure of dispersion that is the square root of the variance. The steps are the same, but you take the square root at the very end. Why is this important? Because it puts the standard deviation in the same units as your dataset and you can apply it to your distribution.

Let's take a step back. The coolest thing about standard deviation is what happens when you apply it to a normal distribution. One standard deviation above the mean will always include 34.1% of your data points as seen in *Figure 7.14* here:

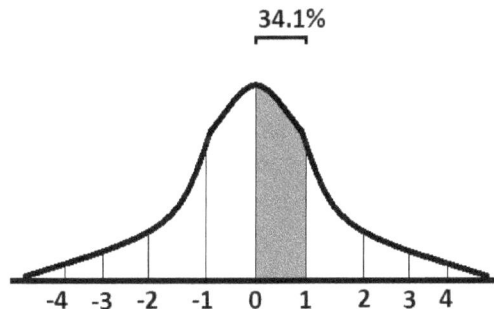

Figure 7.14 – One standard deviation

Because both sides are the same in a normal distribution, that means that one standard deviation below also includes 34.1% of your data points. Together, that means that 68.2% of your data points are within one standard deviation of your mean as seen in *Figure 7.15* here:

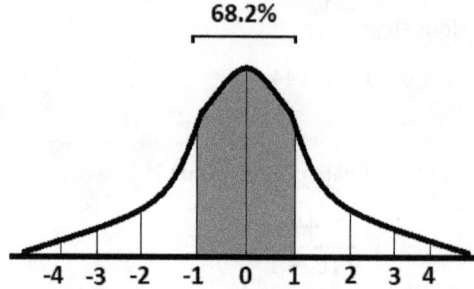

Figure 7.15 – One standard deviation above and below the mean

This keeps going, with 95.4% of all of your data points within two standard deviations and 99.6% of your population within three standard deviations, as seen in *Figure 7.16* here:

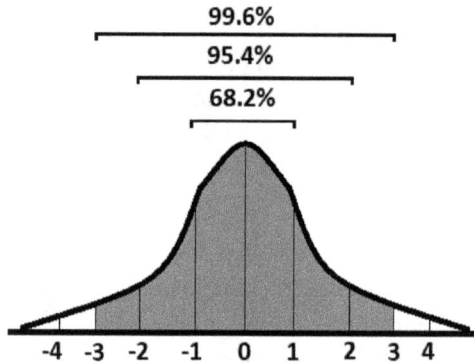

Figure 7.16 – Standard deviation percentages

This is pretty nifty and you can see why many data analysts use three standard deviations as a cutoff for outliers. If you delete every data point that is more than three standard deviations away from your mean, you are only losing 0.4% of your data and removing any outliers while you are at it.

There are many things to learn about standard deviation, but they will not be on the test – so let's go ahead and focus back on what will be. The steps are the same as we had for variance, but we are adding one more to the end:

1. Find the mean of the dataset.
2. Subtract the mean from each data point.
3. Square the results from the previous step.
4. Find the sum of the previous step.
5. Divide that sum by the number of data points minus 1.
6. Take the square root of the previous step.

For our example, we will stick to the boat data, but we will go ahead and collect another sample of 10 days: 4, 5, 3, 4, 2, 2, 6, 4, 6, and 4:

1. Find the mean of the dataset:

$$\frac{(4 + 5 + 3 + 4 + 2 + 2 + 6 + 4 + 6 + 4)}{10} = 4$$

The mean is 4.

2. Subtract the mean from each datapoint:

    $(4 - 4), (5 - 4), (3 - 4), (4 - 4), (2 - 4), (2 - 4), (6 - 4), (4 - 4), (6 - 4),$
    $(4 - 4)$

    becomes

    $(0), (1), (- 1), (0), (-2), (-2), (2), (0), (2), (0)$

3. Square the results from the previous step:

    $0^2, 1^2, -1^2, 0^2, -2^2, -2^2, 2^2, 0^2, 2^2, 0^2$

    becomes

    0, 1, 1, 0, 4, 4, 4, 0, 4, 0

4. Find the sum of the previous step:

    $0 + 1 + 1 + 0 + 4 + 4 + 4 + 0 + 4 + 0 = 18$

    The sum is 18.

5. Divide that sum by the number of data points minus 1:

$$\frac{18}{(10 - 1)} = 2$$

The variance for this dataset is 2.

6. Take the square root of the previous step:

    $\sqrt{2} = 1.41$

    The standard deviation is 1.41. Note that if the square root is not obvious, as here, the answer may be displayed as $\sqrt{2}$.

You can practice this process on your own with the following dataset: 7, 4, 10, 8, 6, 7, 6, 5, 8, 9.

Summary

This chapter covered a lot of information, with a heavy focus on calculations. Just a reminder – while we have spent a fair amount of time learning about these calculations, there will only be a few of these questions on the test. Don't panic, you can get a couple of these wrong and still pass the test if you do well on the other sections.

Distributions show the shape of your data and can tell you a lot about how it is distributed. Common distributions include normal, uniform, Poisson, exponential, Bernoulli, and binomial. Skew is how your data is distorted left or right (negative skew or positive skew). Kurtosis is how your data is distorted up or down (leptokurtic or platykurtic).

Measures of central tendency summarize your data with a single metric. Common forms include the mean, median, and mode. Each measure is best used on specific types of data.

Measures of dispersion include simpler methods such as range and quartiles, as well as slightly more complicated methods such as variance and standard deviation. Ranges, quartiles, and interquartile ranges are all focused on the extent to which your dataset is spread. Variance and standard deviation are more accurate measures that give the average distance of every point from the mean. Standard deviation has an even greater number of uses when applied to a normal distribution because it can break it down into standardized chunks.

If you are not comfortable with arithmetic, you may want to practice the steps covered in this chapter. You can even make up your own datasets and give them a go. The datasets used on the exam will be pretty small and simple and you can use the datasets in the examples of this chapter as a guide.

Practice questions

Let's try to practice the material in this chapter with a few example questions.

Questions

1. The following distribution is considered:

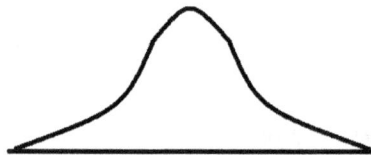

   A. Uniform
   B. Normal
   C. Poisson
   D. Exponential

2. What is the mode(s) of the following dataset?

   24, 18, 36, 51, 24, 48, 18

   A. 18
   B. 24
   C. 18 and 24
   D. None of the above

3. What is the range of the following dataset?

   15, 615, 46, 73, 45, 80, 46

   A. 131
   B. 46
   C. 73
   D. 600

4. What is the middle quartile (Q2) of the following dataset?

   10, 24, 13, 9, 15, 7, 19

   A. 9
   B. 13
   C. 19
   D. 15

5. What is the standard deviation of the following sample dataset?

   9, 11, 7, 8, 9, 10

   A. 2
   B. 9
   C. $\sqrt{2}$
   D. $\sqrt{9}$

Answers

Now, we will briefly go over the answers to the questions. If you got one wrong, make sure to review the topic in this chapter before continuing:

1. The answer is: Normal

    The distribution is the distinctive bell curve of a normal distribution.

2. The answer is: 18 and 24

    Both 18 and 24 are repeated twice, meaning they are both modes.

3. The answer is: 600

    Make sure you identify the minimum and maximum to calculate the range.

4. The answer is: 13

    Remember that the middle quartile (Q2) is the same as the median.

5. The answer is: $\sqrt{2}$

    Make sure to take your time and go through the steps one by one.

Common Techniques in Descriptive Statistics

This chapter has a lot of fun content. This is another chapter of calculations that you will have to know for the exam, so brace yourself and put your math hat on. That said, they are descriptive statistics techniques that are commonly used and good to know. First, we will go over frequencies and percentages. Then, we will talk about how to calculate percent change. Next, we will discuss percent difference, as it is *not* the same thing as percent change. Then, we will cover confidence intervals. Finally, we will go over z-scores and how they are used. Without further ado, let's get started!

In this chapter, we're going to cover the following main topics:

- Understanding frequencies and percentages
- Calculating percent change and percent difference
- Discovering confidence intervals
- Understanding z-scores

Understanding frequencies and percentages

Frequencies and percentages are some of the simplest descriptive statistics, but they are useful nonetheless. This is why they are commonly part of EDA. You get to know the exact makeup of a variable. They can be used in various ways, but are most often applied to a single variable to understand how it is composed. More often than not, this one variable is categorical, so this is a great tool for your qualitative variables.

It should be noted that these can be used for multiple variables at the same time, but how complex they are increases exponentially with every variable added.

Frequencies

We can throw around all the fancy terms we want, but at the end of the day, a frequency is just a count of occurrences. You are literally just counting how many times a specific value occurs within a variable. There are many ways to do this, depending on the data analysis tools you are using, but for this exam, you must calculate it by hand. The easiest tool for this is a frequency table:

Shape	Square	Circle	Triangle
Frequency	42	31	16

A frequency table is exactly what it sounds like: a table that lists every reported value in a variable and how many times they happened. If you recall from the previous chapter, a simplified form of this table was used when finding the mode of a dataset. Let's break this down into easy steps:

1. Arrange the values in ascending or descending order.

2. Create a table and list all possible values.

3. Add a count of every variable.

Let's try this out with an example. The flagship product your company sells comes in an assortment of different colors: Blue, Red, and Yellow. After recording the sales of the product for a day, the results are as follows: Red, Red, Blue, Red, Yellow, Yellow, Blue, Yellow, Red, Red, Blue, Red, Yellow, Blue, and Yellow:

1. Arrange values in ascending or descending order like this:

    Blue, Blue, Blue, Blue, Red, Red, Red, Red, Red, Red, Yellow, Yellow, Yellow, Yellow, Yellow

2. Create a table and list all possible values:

Color	Blue	Red	Yellow
Frequency			

3. Add a count of every variable:

Color	Blue	Red	Yellow
Frequency	4	6	5

You can practice this process on your own with the following dataset: Red, Yellow, Blue, Blue, Yellow, Red, Blue, Yellow, Yellow, Red, Blue, Blue, Yellow, Red, and Blue.

Complex frequency tables

Frequency tables can hold more than one variable and when it does, it is called a contingency table. The steps are the same as making a frequency table for one variable; you simply arrange the table differently.

For two variables, you have one variable on each axis. Let's look at the same product information as before, except now the product also has two sizes: Large and Small. Our dataset now looks like this: Small Red, Small Blue, Large Red, Large Red, Large Yellow, Large Blue, Small Yellow, Small Red, Large Yellow, Small Blue, Small Red, Large Yellow, Large Blue, Large Red, and Small Blue:

1. Arrange values in ascending or descending order:

    Large Blue, Large Blue, Small Blue, Small Blue, Small Blue, Large Red, Large Red, Large Red, Small Red, Small Red, Small Red, Large Yellow, Large Yellow, Large Yellow, Small Yellow

2. Create a table and list all possible values:

	Blue	Red	Yellow
Large			
Small			

3. Add a count of every variable:

	Blue	Red	Yellow
Large	2	3	3
Small	3	3	1

> **Important note**
>
> Do not confuse count with probability. In the preceding example, let's compare Red and Yellow in regard to Large products. The count for Large in each group is 3, but the probability of a Red or Yellow product being large is different. Probability is calculated by dividing the count of that possibility by the sum of counts for that group. If the product is Red, the probability of it being Large is 3/6, or 50%. If the product is Yellow, the probability of it being Large is 3/4, or 75%. In the exam, questions about frequency might be in terms of counts OR probabilities.

Not too bad, right? Okay, let's see what happens when we add a third variable. We are still following the same steps that we did for a frequency table with two variables. This time, we will also track whether the product is Normal or Premium. Let's look at our new dataset: Normal Small Red, Premium Small Red, Normal Large Blue, Premium Small Red, Normal Large Blue, Normal Small Yellow, Premium Large Blue, Premium Small Red, Normal Large Yellow, Premium Large Yellow, Normal Large Blue, Normal Large Yellow, Premium Large Red, Premium Small Blue, and Normal Large Yellow:

1. Arrange values in ascending or descending order:

    Normal Large Blue, Normal Large Blue, Normal Large Blue, Premium Large Blue, Premium Small Blue, Premium Large Red, Normal Small Red, Premium Small Red, Premium Small Red, Premium Small Red, Normal Large Yellow, Normal Large Yellow, Normal Large Yellow, Premium Large Yellow, Normal Small Yellow

2. Create a table and list all possible values:

		Blue	Red	Yellow
Normal	Large			
	Small			
Premium	Large			
	Small			

3. Add a count of every variable:

		Blue	Red	Yellow
Normal	Large	3	0	3
	Small	0	1	1
Premium	Large	1	1	1
	Small	1	3	0

This still may be manageable, but this is also only looking at three variables, and each one of those only has two or three possible values. Imagine 5 variables with 20 values each. It would quickly reach a point where it is difficult to read, time-consuming to put together, and doesn't tell you enough to make it worth it.

Now that you can create a frequency table and scale it to any size, let's move on to percentages.

Percentages

Percentages are really an extension of frequencies. Instead of stopping at counts, they give you the percent of the whole that each value represents. Frequencies do not account for the whole; if you had to visualize them, they would probably be bar charts, as shown in *Figure 8.1*:

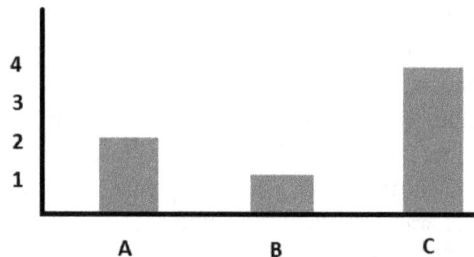

Figure 8.1 – A bar chart symbolizing counts or frequencies

Percentages are all about how much each value impacts the whole, and they would be represented by pie charts, as shown in *Figure 8.2*:

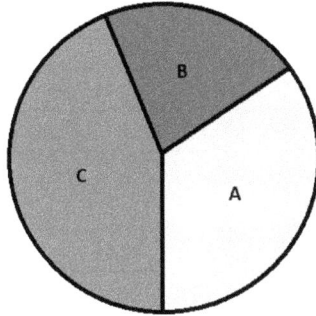

Figure 8.2 – A pie chart symbolizing percentages

If you are unfamiliar with the specifics of these charts, we will cover them in detail in *Chapter 13, Common Visualizations*.

The steps to find percentages are straightforward:

1. Find the frequencies.
2. Divide each frequency by the total number of values.
3. Multiply each frequency by 100.

Let's try these out in an example. Let's keep the previous example of a product, but for the sake of simplicity, we will only be looking at the color variable. The dataset is as follows: Red, Red, Yellow, Red, Blue, Blue, Yellow, Red, Blue, and Yellow:

1. Find the frequencies.

    We just went through how to do this. If you are unsure, please review the previous section:

Blue	Red	Yellow
3	4	3

2. Divide each frequency by the total number of values.

    Our total number of values, reached by adding together all of the counts, is 10:

Blue	Red	Yellow
3/10	4/10	3/10

    This gives us the following:

Blue	Red	Yellow
0.3	0.4	0.3

3.   Multiply each frequency by 100:

Blue	Red	Yellow
0.3*100	0.4*100	0.3*100

This gives us the following:

Blue	Red	Yellow
30%	40%	30%

This gives us the percentages of each value that makes up the whole of our sample. Percentages are useful and are heavily used with demographic data. If 70% of your customers fall into a specific age bracket, you will want to target your advertisements toward that age bracket. There is a wide range of key demographics, but the majority of them are qualitative data that benefit from a quick percentage table.

Now that we have introduced the idea of percentages, let's talk about percent change.

Calculating percent change and percent difference

Percent change and percent difference are very similar concepts, but they have slightly different equations. In the exam, you will be expected to not only be able to calculate each but also understand when to use each. Be careful, the exam may have the correct answers to both equations as possible choices, so you can't just calculate both and see which is on the list of answers.

Percent change

When thinking about percent change, the word you need to focus on is **change**. The idea is straightforward; you are concerned with how a single value has changed from point A to point B. You are just reporting this change as a percent, in reference to the starting value. When we say "in reference to a starting value," we are trying to communicate that the starting value is the star of the show. It is subjective; all of this is told in terms of the starting value. Does the value increase or decrease since the starting value? Let's consider an example.

What is the percent change in the ROI from the beginning of the year until now? This can be read as "How much has the ROI changed since the beginning of the year?" You do the calculations and find the answer, "The ROI has increased by 5% since the beginning of the year." Here, we are identifying the single metric (ROI), how it has changed (increased), how much it has changed as a percent (5%), and the starting point of reference (since the beginning of the year).

> **Important note**
>
> It should be noted that you can have positive or negative results from a percent change equation. Positive results show an increase in the value (the end point is larger than the starting point) and negative results show a decrease in the value (the end point is smaller than the starting point).

In the exam, the answer choices will probably not be laid out this cleanly, but this example illustrates the concept. Again, percent change is all about how a single value changes in reference to a starting value. You will always have a variable with a starting value and an ending value, so pay attention to which is which. Let's look at the equation to understand why:

$$C = (\frac{x_2 - x_1}{x_1}) * 100$$

In this equation, C is the percent change, x_1 is the starting value, and x_2 is the ending value. Getting the starting and ending values mixed up will give you an incorrect answer. The actual calculations are not that bad on this one, so let's look at the steps:

1. Subtract the starting value from the ending value.
2. Divide the results of the previous step by the starting value.
3. Multiply the results of the previous step by 100.

For our example, we will be looking at the metric *clicks per minute*, which describes the average number of times a web advertisement is clicked every minute. At the beginning of 2021, the clicks per minute was 4, and by the end of the year, it was 6. Let's see how much the clicks per minute changed:

1. Subtract the starting value from the ending value.

   We can identify the starting value as 4 and the ending value as 6:

   6 – 4 = 2

2. Divide the results of the previous step by the starting value:

   2 / 4 = 0.5

3. Multiply the results of the previous step by 100:

   0.5 * 100 = 50%

   The percent change is 50%, which means the clicks per minute has increased by 50% since the start of the year.

To practice this process on your own, try using the following values: starting value = 1 and ending value = 3.

Now that you have mastered percent change, it is time to move on to percent difference and discuss how they are different.

Percent difference

The percent difference is like percent change; they both compare two values using percentages. Where they differ is that with a percent difference, the two values have equal weight, and you are no longer giving priority or writing in reference to a starting point. The two values can be from completely different sources. Instead of looking at how your ROI has changed over the year, you can now look at how your ROI compares to your competitor's ROI. How do the sales of this department compare to the sales of that department?

The tricky thing is that this can still apply to the change in a single value, but there will be something that draws a line between the two values. This seems like a subtle difference, but it changes the values from "starting and ending" to "before and after." For example, you get a job working as a process analyst at a manufacturing plant. You look at the maintenance cost per unit to keep the machines working. At the start of the year, this value was $200. Six months later, the machines are replaced with a more efficient model, and the maintenance cost per unit is now $100. It is the same metric, and we are looking at it at the start of the year and halfway through, but what we really want to know is how the value before the new machines compares to the value after the new machines. You can think of this as artificially creating two different sources.

What we are really doing is looking at the difference between two different numbers to compare relative sizes. Let's look at the equation:

$$\text{Percent difference} = \left(\frac{|x_1 - x_2|}{\left(\frac{x_1 + x_2}{2} \right)} \right) * 100$$

If you look carefully at this equation, you will realize that it doesn't matter which value is x_1 and which value is x_2. Because you are taking the absolute value, you cannot have a negative answer, so $|2 - 1| = 1$ but $|1 - 2| = 1$ as well.

> **Important note**
>
> Because of this, unlike percent change, percent difference will never have a negative value. There is no "increasing" or "decreasing." You are only looking at the objective difference between two values.

The steps are as follows:

1. Subtract one value from the other and drop any negative signs.
2. Find the average of the two values.
3. Divide the result of the first step by the result of the second step.
4. Multiply the result of the previous step by 100.

For consistency, let's use the same example as we did with percent change, but this time, the results come from two different ad campaigns. The clicks per minute from ad campaign A were 4 and from ad campaign B were 6. What is the percent difference?

1. Subtract one value from the other and drop any negative signs:

    4 – 6 = -2

    If we drop the negative sign, that changes the result:

    2

2. Find the average of the two values:

    (4 + 6) / 2 = 5

3. Divide the result of the first step by the result of the second step:

    2 / 5 = 0.4

4. Multiply the result of the previous step by 100:

    0.4 * 100 = 40%

    The percent difference is 40%.

You will notice that this is not the same value as the percent change using the same values. That is because it is not in terms of the starting value. This is the objective difference between two values of equal importance.

To practice this process on your own, try using the following values: starting value = 1 and ending value = 3.

We have covered percentages of one sort or another plenty. Let's move on to another common tool that is often argued over: the confidence interval.

Discovering confidence intervals

Confidence intervals are used in different ways depending on the field, but the basic idea is the same. A confidence interval is a range including the mean of your sample. We see what this looks like in *Figure 8.3*:

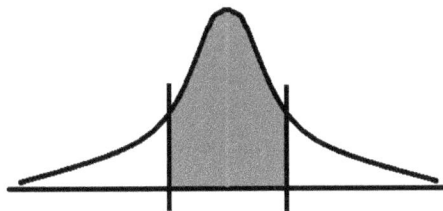

Figure 8.3 – Confidence interval

What does this interval actually mean? Some say that you are confident that the mean of the population, or "true mean," will fall within this range, some say that you are confident that if you repeat the study, the new mean will fall within this range, and others say that you are confident that the actual value of an estimate will fall within this range. This is a heated and completely pointless debate; they effectively say the same thing. That said, for the purpose of this exam, we will assume the first definition is correct: you are confident the "true mean" falls within the interval.

How confident you are is really defined by the confidence level, which is a percent value that represents the probability of you being correct. Technically, it is 1 minus your alpha times 100. We will discuss alpha in detail in *Chapter 9, Hypothesis Testing*, but for now, know that the most common alpha is 0.05, which gives you a 95% confidence level. A confidence interval with a confidence level of 95% is written as a 95% confidence interval. This means that you are 95% sure that the mean of the population falls within this range.

The equation for the confidence interval is as follows:

$$CI = \bar{x} \pm t \left(\frac{\sigma}{\sqrt{n}}\right)$$

$\bar{x}$ is the mean, t is a value from a t-distribution confidence table, σ is your standard deviation, and n is your sample size. This equation creates a range by generating two values: the lower confidence interval and the upper confidence interval. These two values together make a range around the mean.

The value for t is looked up on a table using the alpha and the **degrees of freedom (dfs)**. We briefly discussed alpha, but the **df** is simply your sample size minus 1. You may want to try looking up t-distribution confidence tables and figuring out values for t in different situations for fun, but it's not really a skill you need. For the exam, the value of t should be provided, and almost every data analytics tool will just calculate these for you.

Enough talking, let's go ahead and look at the steps for calculating confidence intervals:

1. Find the mean of your sample.
2. Find the standard deviation of your sample.
3. Find the square root of your sample size.
4. Divide your standard deviation by the result of the previous step.
5. Multiply the result of the previous step by your t-value.
6. Add the result of the previous step to your mean to find your upper confidence interval.
7. Subtract the result of *step 5* from the mean to find your lower confidence interval.

This has a few steps, so let's run through an example. You are hired as the main data analyst for a hotdog cart. They want to know how many hotdogs they sell per hour. You recorded how many hotdogs they sold per hour for several hours and created the following dataset: 8, 6, 9, 7, 8, 10, 8, 8, 8. The t-value for this sample is 1.86:

1. Find the mean of your sample.

   The mean is 8.

2. Find the standard deviation of your sample.

   The standard deviation is 1.12.

3. Find the square root of your sample size.

   The square root of the sample size is 3.

4. Divide your standard deviation by the result of the previous step:

   1.12 / 3 = 0.37

5. Multiply the result of the previous step by your t-value:

   1.86 * 0.37 = 0.69

6. Add the result of the previous step to your mean to find your upper confidence interval:

   Upper confidence interval = 8 + 0.69 = 8.69

7. Subtract the result of *step 5* from the mean to find your lower confidence interval:

   Lower confidence interval = 8 – 0.69 = 7.31

   We are 95% confident that the "true mean" of hotdogs sold per hour is between 7.31 and 8.69.

To practice this process on your own, try using the following dataset: 2, 1, 3, 2. The t-value is 2.35.

Next, we will move on to a value we briefly discussed in a previous chapter, the z-score.

Understanding z-scores

Z-scores, sometimes called standard scores, have a very specific purpose, and for once, no one is debating it. The purpose of a z-score is to compare a single value to a normal distribution. More specifically, a z-score reports how many standard deviations your chosen value is from the mean.

> **Important note**
> You will not be asked to calculate a z-score for the exam. You will simply have to understand the purpose of it and identify when it is appropriate to use. That said, it is simpler than a lot of the things you have already learned to calculate, and practicing with it will help reinforce it in your memory.

This has many uses. Z-scores are often used for things such as percentiles, but let's take a broader view. Any time you want to compare a single value to a dataset, the z-score is a good choice. Is this price too high? Is this weight too low? How do my grades compare to the other applicants? Can we market to this age bracket?

There are countless fun things you can figure out, so let's jump right in and talk about the equation:

$$Z = \frac{(x - \mu)}{\sigma}$$

Easy enough. Here, Z is your z-score, x is the value you want to compare, μ is the mean of your distribution, and σ is the standard deviation.

The steps are as follows:

1. Subtract the mean from your value.
2. Divide the result of the previous step by the standard deviation.
3. Rejoice!

Let's go ahead and look at an example. Your distribution has a mean of 5 and a standard deviation of 1. How does the value of 6 compare?

1. Subtract the mean from your value:

    6 – 5 = 1

2. Divide the results of the previous step by the standard deviation:

    1 / 1 = 1

    The value 6 is 1 standard deviation above the mean. We can see this in *Figure 8.4*:

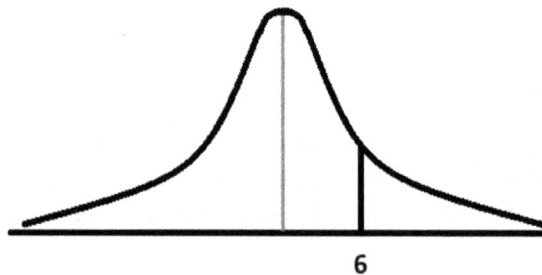

Figure 8.4 – Distribution with result marked

If you recall, in a normal distribution, each standard deviation accounts for a specific percentage of the whole. A value that is 1 standard deviation above the mean is higher than roughly 84% of all values and is lower than only roughly 15.8% of values. We can see this illustrated in *Figure 8.5*:

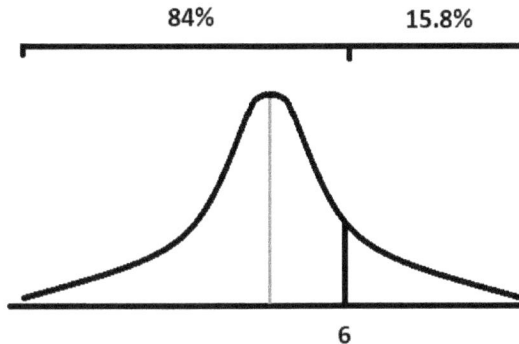

Figure 8.5 – Distribution with result and percentages

3. Rejoice!

Yay! You have survived the two chapters of hand calculations!

Summary

This has been a fun chapter and it covered a lot of useful tools. First, we discussed frequencies, which are counts of every possible value, and how to create frequency tables. Then, we discussed percentages as an extension of frequency that describes how the counts affect the whole of the data.

We talked about percent change, which is how a single value changes. It will always have a starting value and an ending value and be told in terms of the starting value. It matters which value is which. With percent change, you can have positive and negative values that correspond to the value increasing or decreasing. Percent difference is not quite the same thing. It describes the objective difference between two values of equal importance. As such, it does not matter which value is which, and it cannot have a negative value.

Confidence intervals describe a range around the mean of your sample distribution. You are confident that the "true mean" of your population falls within this range. How confident you are is determined by your confidence level, where 95% is the most common.

Z-scores allow you to compare a single value to a distribution in terms of standard deviation.

Just like that, you have finished learning about all of the calculations you will need to know how to work out by hand. These can be tricky, so make sure you practice! Next, we will be getting into the key concepts behind hypothesis testing!

Practice questions

Let's try to practice the material in this chapter with a few example questions.

Questions

1. Convenient News, a membership service, has created the following frequency table about its subscribers. The table describes the sex of the member, whether they were a Normal or Premium member, and whether they continued their subscription or quit. Which group described in the table has the highest probability of quitting?

		Current Subscriber	Past Subscriber
Male	Normal	89	52
	Premium	12	2
Female	Normal	12	45
	Premium	35	13

    A. Males with a Normal subscription

    B. Males with a Premium subscription

    C. Females with a Normal subscription

    D. Females with a Premium subscription

2. Convenient News has tracked the **Customer Lifetime Value** (**CLV**), a common KPI, for a while now. At the start of the year, the CLV was $200, and now it is $240. How do these numbers compare?

    A. 20% increase

    B. 20% decrease

    C. -18% difference

    D. 18% difference

3. Convenient News wants to compare its CLV of $240 to a competitor, Inconvenient News, which has a CLV of $180. How do they compare, in percentages?

    A. -28% difference

    B. 28% difference

    C. 25% increase

    D. 25% decrease

4. Convenient News is tracking how many newspapers a specific store is selling every day. The dataset is as follows: 10, 12, 11, 8, 10, 8, 9, 12, 10. The standard deviation is 1.5, the t-value is 1.86, and the mean is 10. Create a confidence interval for this data:

    A. The lower confidence interval is 9.75 and the upper confidence interval is 10.25

    B. The lower confidence interval is 8.14 and the upper confidence interval is 11.86

    C. The lower confidence interval is 8.5 and the upper confidence interval is 11.5

    D. The lower confidence interval is 9.07 and the upper confidence interval is 10.93

5. Convenient News has tracked the number of times a particular article has been read on its website for years. A specific article stands out and the author wants to know how the number of times it has been read compares to other articles. Which analysis is most appropriate?

    A. Simple linear regression

    B. T-test

    C. Z-score

    D. Chi-square

Answers

Now, we will briefly go over the answers to the questions. If you got one wrong, make sure to review the topic in this chapter before continuing:

1. The answer is: Females with a Normal subscription

    Females with a Normal subscription is the only group that has had more members quit than stay. The probability of this group quitting is 79%. Even though the count of Men with Normal subscriptions who quit is higher, the probability of this group quitting is only 37%, because there are so many more members in this group.

2. The answer is: 20% increase

    This is a percent change question because you are looking at a single value at two different points in terms of the starting value.

3. The answer is: 28% difference

    This is a percent difference problem because you are looking at two distinct groups. Also, remember that percent difference cannot be a negative value.

4. The answer is: The lower confidence interval is 9.07 and upper confidence interval is 10.93

    You are given all of the pieces for this one, it is just a matter of going through the steps one at a time.

5. Z-score

    Z-scores are specifically for comparing a single value to a distribution. Here, you are comparing the value of a metric for a single article against all past values for every other article.

9
Hypothesis Testing

This chapter is all about the basic concepts behind hypothesis testing. Some data analyst positions will have little to no hypothesis testing, but this is really where data analysts shine. You receive a question, create a hypothesis, sometimes run a study, analyze data, and give a report that allows someone to make an informed decision. This process is the heart of data analytics, even if the majority of your time will be spent on more mundane tasks.

Here, we will discuss what hypothesis testing is and how it relates to the null hypothesis and the alternative hypothesis. Then, we will cover p-value and alpha and their role in hypothesis testing. Next, we will discuss type I and type II errors, how you can adjust them, and how they impact your hypothesis testing. Finally, we will go over how to write a question that will lead to a clean and accurate hypothesis test.

In this chapter, we're going to cover the following main topics:

- Understanding hypothesis testing
- Differentiating null hypothesis and alternative hypothesis
- Learning about p-value and alpha
- Understanding type I and type II errors
- Writing the right questions

Understanding hypothesis testing

In this section, we will cover what hypothesis testing is and why it is an important skill to have for a data analyst. Hypothesis testing is, in a nutshell, trying to guess whether or not your hypothesis is true. If someone asks you a specific business question, this is how you give a confident yes or no answer.

There are several different types of hypothesis tests, as you can imagine, but, for the purpose of this chapter, we are going to focus on one of the most common ones. The most common form of hypothesis testing is comparing two things and saying whether or not they are the same. It is as simple as that. Sometimes, you are comparing two completely separate things. Sometimes, you are comparing a "before" and "after" group. Often, you are comparing something altered to something unaltered. Which product sells better, A or B? Are our sales this year higher than last year? Would we sell the product to more women if we put an adorable mascot on the front?

Why use hypothesis testing

Okay, it's easy to see why answering these questions is important, but why go through hypothesis testing instead of just looking at the numbers? There are a few things that come into play here. Usually, you are comparing groups, not individual numbers. More importantly, even if you compare two means, the question becomes is the perceived difference between the groups just due to variance, or is one actually higher? In *Figure 9.1*, we can see how comparing two distributions can be an iffy prospect.

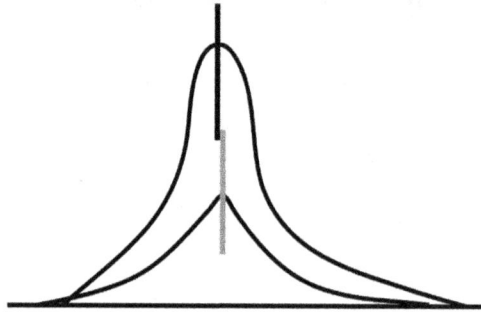

Figure 9.1 – Comparing two distributions

Let's consider an example. You have two groups: A and B. The mean of A is 100 and the mean of B is 101. Group B is higher, right? Well, what if both groups had a lot of variance? Let's say the standard deviation of group A is around 60. That means that 68% of group A's distribution is somewhere between 40 and 160. Alternatively, what if group B was a tiny sample compared to group A? Could you confidently say that it was not just a random chance that the mean of B was 1 point higher? Maybe if you had a bigger sample, they would even out, or group A might be higher in the long run. In *Figure 9.2*, we can get an idea of what these numbers might look like.

Group A	Group B				Mean Standard Deviation
110	26		Group A	100	63.2455532
10	178		Group B	101	76.01973428
90	99				
190					

Group A	Group B				Mean Standard Deviation
180					
20					
50					
150					
130					
70					

Figure 9.2 – Example of two datasets being compared

Being able to say, with a very small margin of error, that there is an actual difference beyond pure chance between the two groups is called statistical significance. That is our goal here. Is there a statistically significant difference between the two groups or not?

This is a process, and not just a single analysis. You need to go through step by step and understand the decisions you are making and why you are making them. If you just run an analysis and hope for the best, then you may guess correctly, but you might also have tested the wrong thing altogether, or your results might be the opposite of what you thought because of how things are arranged.

Hypothesis testing process

The process of hypothesis testing can vary, depending on your field and what you are doing, but these general steps are true more often than not:

1. Ask or receive a question.
2. Create a hypothesis.
3. Decide what analysis is most appropriate to answer your question.
4. Gather the resources required for analysis.
5. Run the analysis.
6. Answer the question.

Rinse. Repeat. Let's go over this at a high level:

1. Ask or receive a question.

    First, everything starts with a question. It doesn't matter whether your boss asks it or you found something odd in the data while you were fiddling around and asked it yourself. It doesn't matter what it is – there will be questions – and it is the job of a data analyst to use the data to try to answer these questions and give insights, so that others may make informed decisions. For example, do Jim's muffins taste better than mine?

2.   Create a hypothesis.

Creating a hypothesis is simply restating the question as a testable statement with a prediction of the outcome. A good question makes this easy and will translate directly, but you may also have large or ambiguous questions that will require several hypotheses to answer. Our example might be, "Jim's muffins will have a higher approval rating than mine."

3.   Decide what analysis is more appropriate to answer your question.

Picking an analysis is an important step, not only because you are deciding how to answer the question but also because you can identify everything you need to answer the question. Here, you identify what variables you need, how many data points you need, and anything else that might be an assumption of your analysis. Maybe a t-test is the best way to decide whose muffins are better, so you need a set of muffin ratings for yourself and Jim, as well as all assumptions required for this test. If you don't know what a t-test is, it will be covered in *Chapter 10, Introduction to Inferential Statistics*.

4.   Gather resources required for analysis.

Gathering the resources is just about going around and collecting the things you identified in the previous step. Sometimes, this is as simple as running a query, and sometimes it is as complicated as designing and running an entire study to collect new information. This is where you go around the office, force-feeding people muffins and asking them to give each a rating.

5.   Run the analysis.

Running the analysis is, arguably, the easiest step. Often, this is a matter of clicking a few buttons or reworking a code example you have used in the past to pull from the correct sources. A lot of the data analytics tools are designed to make this as simple as possible. Here, we would run the t-test on the muffin data.

6.   Answer the question.

Finally, answer the question. This is interpreting the results and reporting your findings. Yes, your hypothesis was correct, Jim's muffins did have a higher approval rating. Maybe it's because he uses fresh blueberries, or maybe it was just that people rated your muffins lower because you were bullying them into taking a survey when they were busy.

Differentiating null hypothesis and alternative hypothesis

This is the point where a lot of people get confused. When you are testing a hypothesis, you are not directly testing the hypothesis statement you write out at the beginning. You are testing to see whether the null hypothesis or the alternative hypothesis is true. Whatever you wrote out as your hypothesis will be linked to one of these two options. In other words, you are either agreeing with the null hypothesis or the alternative hypothesis. If you are still a little confused, it will make sense when we get into what these two options are.

Null hypothesis (H_0)

The null hypothesis, most often denoted as H_0, is the default hypothesis and says that there is no difference between the two groups. Being the default means that the two groups are the same until proven otherwise.

Hypothesis	Conclusion
Null	There is NO significant difference
Alternative	

This is the hurdle that must be overcome to prove that the groups are statistically different.

Alternative hypothesis (H_1)

The alternative hypothesis, most often denoted as H_1, says (you guessed it) that the two groups are different. This is simply a statement that the two groups, whatever you are comparing, have a statistically significant difference between them.

Hypothesis	Conclusion
Null	There is NO significant difference
Alternative	There IS a significant difference

It doesn't tell you which group is higher or lower – that you have to check on your own. It only tells you that they are not the same.

Null hypothesis versus alternative hypothesis

The null hypothesis and alternative hypothesis are two sides of the same coin. That means that they are mutually exclusive. To be explicit, this means that one is always true and the other is always false. There are no circumstances where both are true, or both are false; hypothesis testing has no room for gray areas. If a hypothesis is true, you accept it and reject the other one.

All Possible Results
accept H_0 and H_1 reject
accept H_1 and H_0 reject

There are only two possible outcomes to any hypothesis test. Either you will accept H_0 and reject H_1 or you will accept H_1 and reject H_0. A lot of people will focus on either H_0 or H_1 and only report in terms of that, but saying both and making it clear which one is being accepted and which one is being rejected can clear up a lot of confusion. This is because data analytics is made up of so many fields and each has a different perspective, so it is best to be as clear as possible. For example, in the field of biology, things are often discussed in the terms of disproving the null hypothesis, while in the field of psychology, things are more often discussed in terms of proving the alternative hypothesis. These are the same thing, but it makes communication strangely difficult until everyone gets on the same page. Clearly stating what you accept and what you reject removes any room for confusion.

Let's look back at the muffin study. We wrote our hypothesis statement as "Jim's muffins will have a higher approval rating than mine." Two conditions must be met for this to be considered true. First, the outcome of the analysis must lead us to accept H_1 and H_0 reject . This means that there must be a statistically significant difference between the ratings of Jim's muffins and the ratings of your muffins. The second thing that must be true is that the mean of the ratings of Jim's muffins must be higher than the mean of the ratings of your muffins. Remember, the actual analysis only told you that there was a difference; you still need to check to see what that difference is. Theoretically, there could be a difference, but your mean was higher, which would still prove your hypothesis statement was wrong. The actual analysis will only lead you to H_0 or H_1 ; it is up to you to connect the dots between this and your hypothesis statement.

Okay, you get the idea of H_1 and H_0, but how do you actually decide which to accept and which to reject? Well, that has to do with p-value and alpha.

Learning about p-value and alpha

Let's talk about the results of an analysis and how those translate to accepting and rejecting hypotheses. First, it should be noted that many different analyses can be used in hypothesis testing, and a few might give their results in different terms. However, the majority of tests that are used for hypothesis testing produce something called a p-value. This value is either above or below a cutoff score, set by the alpha, and that determines which hypothesis you accept and which you reject. Let's look at these topics in more detail.

p-value

A p-value is a number produced by many analyses used in hypothesis testing. A p-value can, theoretically, be any value between 0 and 1. This may sound like a small range, but with decimals, there is literally an infinite number of values between 0 and 1. If you get into the technical definition of the p-value, it gets needlessly complicated very quickly. You can think of the p-value as the probability that the difference between the groups is a fluke. This means that if you get a p-value of 0.05, there is only a 5% chance that the difference between the two groups was just random. We will discuss alpha in just a moment, but most people would accept this 5% risk of being wrong, accept H_1, and reject H_0.

Alpha

The alpha, sometimes called the significance level, determines the cutoff or threshold for statistical significance. What does that mean?

- p-value ≤ alpha = Accept H_1 and reject H_0
- p-value > alpha = Accept H_0 and reject H_1

If your p-value is equal to or smaller than your alpha, you accept H_1 and reject H_0. If your p-value is greater than your alpha, you accept H_0 and reject H_1. In other words, if your probability of being wrong is below your target threshold, you accept that the difference between the groups is significant and not just down to chance. There is a difference!

If, however, the probability of being wrong is too high and is above the line of what you think is acceptable, then you just have to admit that the difference between the groups is probably just down to chance. There is no significant difference.

Let's look back at our muffin example. You gather the muffin ratings and compare them with appropriate analysis. The test gives you a p-value of 0.042. You go with the default alpha of 0.05. What does this mean?

This means that the probability that the perceived difference between the two groups is just down to chance is 4.2%. More importantly, the p-value is smaller than the alpha (0.042 < 0.05). We accept H_1 and reject H_0. Let's put this into words one more time. The probability that the difference between the groups is just random chance is low enough that we accept the alternative hypothesis (there is a difference) and reject the null hypothesis (they are the same).

Alpha and tails

When you are performing a hypothesis test, you can choose whether it is one-tailed or two-tailed. This has to do with how your alpha is distributed. By default, most people and analytical tools run two-tailed tests. That means that if your alpha is 0.05, half of it is at the far left of the distribution, and half of it is at the far right. We can see this in *Figure 9.3*:

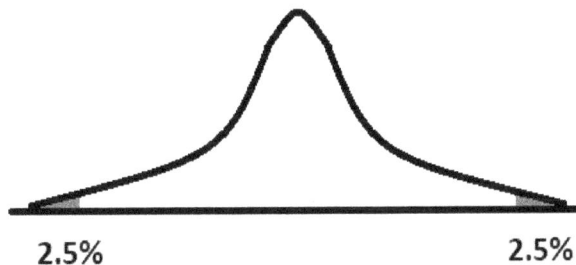

2.5%   2.5%

Figure 9.3 – Two-tailed

This way, you know that there is a difference between the two groups, and it doesn't matter which is bigger and which is smaller. However, there is a specific kind of analysis called one-tailed, which puts all of the alpha on one side or the other. This is illustrated in *Figure 9.4*:

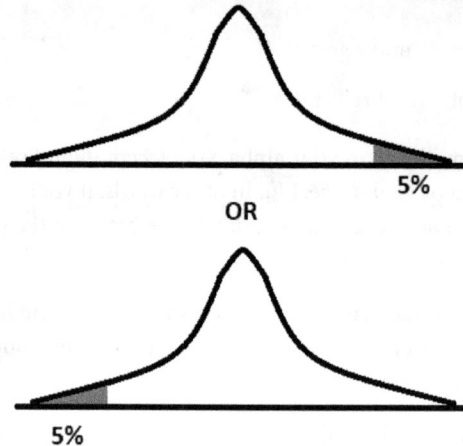

Figure 9.4 – One-tailed

One-tailed tests are pretty rare in the industry, but they can be used if you only care that the groups are different in a very specific way. This book, like most industry tools, will assume all hypothesis tests are two-tailed unless specifically noted otherwise.

Okay, there are a few things we need to cover before we move on from alpha. First, all of this is based on probability, so there is always some chance of being wrong no matter what you do. Our goal is to minimize that chance. Second, you may remember alpha from when we discuss confidence. You can flip this on its head and say that a 5% chance of being wrong is a 95% chance of being right! This, again, links back to confidence levels.

Finally, you can set your alpha to anything you like, but, in practice, there are only three different values of alpha that you will come across: 0.1, 0.05, and 0.01. These correspond to a 10%, 5%, and 1% chance of saying the groups are different when they are actually the same. When in doubt, just use 0.05, because it is considered a balance, while 0.1 and 0.01 are considered a little extreme. To understand why that is and when to adjust your alpha, we need to talk about type I and type II errors.

Understanding type I and type II errors

When we are talking about errors in hypothesis testing, we are not talking about typos, or even mistakes made with the study to gather data. We are talking about coming up with the wrong answer. Earlier, we went over there only being two outcomes to a hypothesis test; either you accept H_1 and reject H_0 or you accept H_0 and reject H_1. Those are the only two possible outcomes. We also discussed that this is based on probability and there will always be a small chance that you choose incorrectly. Even a

95% chance of being right means you are wrong 1 in 20 times. In other words, both of the two possible outcomes have an error associated with them. These errors are called type I and type II errors.

Type I error

Type I error is when you incorrectly accept H_1 and reject H_0. Your p-value was smaller than your alpha, so you said there was a statistically significant difference between the two groups, but it turns out they were really just the same thing. Going back to our muffin example, the test gave us a p-value of 0.042, which was smaller than our alpha of 0.05, so we accepted H_1 and rejected H_0, but that 4.2% chance came true! There wasn't actually a difference in the muffin ratings at all.

Error	Cause
Type I	Falsely accept alternative hypothesis and reject null hypothesis
Type II	

Now, if we are being picky, alpha is specifically the probability of a type I error. When we use an alpha of 0.05, it means we are accepting a 5% chance of a type I error, or we will say the groups are different when they aren't.

Sometimes, this is referred to as a "false positive," and you see lots of images of men being told that they are pregnant when they can't be. The problem with these charts is that half of them are from the perspective of the null hypothesis and half of them are from the perspective of the alternative hypothesis, so all of the values are flipped and they don't always say what their perspective is. This leads to a lot of people being confused. While it is not very catchy, you are better off trying to remember that type I error is when you say there is a difference, when there wasn't one.

Type II error

Type II error is when you incorrectly accept H_0 and reject H_1. Your p-value was larger than your alpha, so you said there was no difference between the groups, but it turns out there was. This is the exact opposite of a type I error and there are only two choices, so you only have to memorize one of the error types and you will always be able to figure out the other one.

Error	Cause
Type I	Falsely accept alternative hypothesis and reject null hypothesis
Type II	Falsely accept null hypothesis and reject alternative hypothesis

If the alpha describes the probability of type I error, then the beta tells you the probability of type II error. We are not going to go into detail about beta, because it is a deep statistical rabbit hole. More importantly, beta will not be on the exam. However, it isn't bad to know that a variable called beta exists, is linked to the power of your analysis, and is related to your alpha.

How type I and type II errors interact with alpha

Here is the rub. The reason why there are different levels of alpha, and why most people don't just use the smallest one for everything, is that as alpha decreases beta increases and vice versa. That means as the probability of type I error goes down, the probability of type II error goes up! Your net chances of being wrong don't actually change that much; you are actually changing the probability of being wrong in a specific way.

Generally speaking, an alpha of 0.05 is a balance of type I and type II errors that lands somewhere in the middle. This is why it is such a common value. Again, when in doubt, an alpha of 0.05 will work. We can see this in *Figure 9.5*:

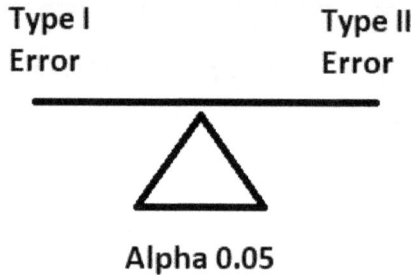

Figure 9.5 – Balance of type I and type II errors with alpha as 0.05

An alpha of 0.01 is considered a rigorous approach. The probability of a type I error is very low, but the probability of a type II error is much higher. In other words, you are very likely to say there is no difference when there actually is, but you only say there is a difference if there is a big difference. This approach of saying the groups are different when they really aren't comes at a high cost. If you are running an analysis to see whether a man is innocent or should be executed, you only want to accept that alternative hypothesis if you are *very* sure. This is illustrated in *Figure 9.6*:

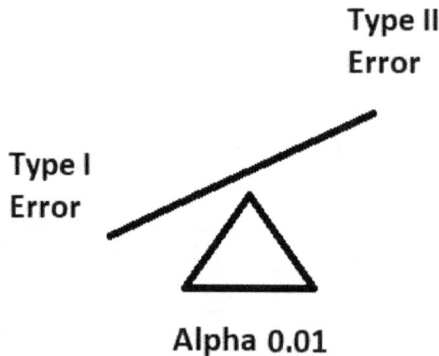

Figure 9.6 – Balance of type I and type II errors with alpha as 0.01

An alpha of 0.1 is the opposite. You have a much higher chance of a type I error, but a lower chance of a type II error. This is usually used when there are few to no downsides to saying something is different when it isn't. Say you are running an analysis to see whether someone would benefit from a high-five. There is no problem with giving a high-five to someone who doesn't need it, but not giving someone a high-five when they are depressed is an issue. Let's look at this in *Figure 9.7*:

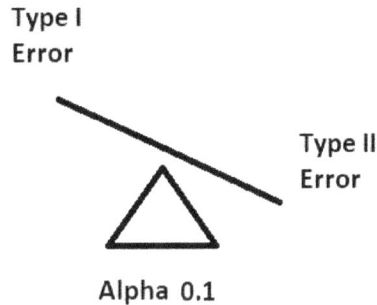

Figure 9.7 – Balance of type I and type II errors with alpha as 0.1

Let's take a breather and step away from statistics for a moment. From a purely practical standpoint, this entire process starts with a question, but what makes a good question for hypothesis testing?

Writing the right questions

Here, we will go over what makes a good question when it comes to hypothesis testing. This question is called many different things, depending on the field: business question, research question, or even just question. However, these are all the same thing. The question is the reason for the hypothesis testing! This question is what you are trying to answer.

> **Important note**
>
> The specifics of what makes a good question might change a little based on the field. For academia, you may want a question that will take several studies and a dissertation to answer, but for this book, we are focusing on the common role of data analysts working in the industry.

The parts of a good question

A good question has only two parts:

- What two groups you want compared
- The metric you want to use to compare those groups

Let's think of an example. Does product X sell more per week than product Y?

We want to compare product X to product Y – those are our two groups. The number of sales per week is the metric we will use to compare them.

It doesn't matter how the question is arranged as long as you have those basic elements.

How do product X and product Y compare on sales per week?

Which product has more sales per week, product X or product Y?

As long as you know what groups you are comparing and how you are comparing them, you will be in a good position to start your hypothesis testing.

Qualities of a good question

This is a little more flexible than the actual parts of a good question, but they will generally have the same qualities:

- The question is short
- The question is easy to understand
- The question is specific
- The question is something you can answer

It is easy to see why long questions or complicated questions are a problem, but the biggest problem you will run into is how specific a question is. Vague, ambiguous questions are the bane of data analysts everywhere. Things such as, "How are we doing this year?", "Are we beating the competition?", or "What is the meaning of it all?" are examples of questions that are difficult to answer.

When we are talking about whether the question is something you can answer, we are talking about the data available to you. If the question wants to use a variable that you have never collected before to compare last year's data, then you may be in trouble.

What to do about bad questions

Sometimes, you receive a question from your boss, and you have to answer it, even if you would rather not. The first thing you should always do is ask for clarification. This may sound obvious, but it is something a lot of analysts overlook or feel awkward doing. It's okay. Sit down with the person who asked the question and try to figure out specifically what they want to know and why. Sometimes, the question they give you is not the question they need answering. They will appreciate getting the answers they need, even if they have to sit down and explain it for a few minutes. It is better than spending a week trying to answer a question and then having them tell you that wasn't what they wanted to know.

If for whatever reason this is not an option, then you need to break the question down into little questions that you can answer. If you are asked "How are we doing this year?", you can break it down into "How do our KPIs this year compare to last year?", "How do our KPIs this year compare to our projections for this year?", "How do our KPIs this year compare to the metric goals we set at the beginning of the year?", and "How do our KPIs this year compare to the KPIs of our major competitors?" Sit down and come up with a list of smaller questions that you can answer, and combine them to answer a large, vague question.

> **Important note**
>
> When reporting results, it is best practice to answer each smaller question one at a time. While it may be tempting to answer all four of your smaller questions with a single, massive visualization, please resist the urge. Each visualization should tell a single, clear, and understandable message. This is your data story. Trying to answer lots of questions with one chart may seem like an efficient use of your time, but chances are you will spend even more time explaining it, or, more likely, the audience will just misunderstand and never ask for clarification.

Summary

In this chapter, we talked all about hypothesis testing. We covered what it was, as well as a quick guide to the process of hypothesis testing. Next, we discussed a null hypothesis, an alternative hypothesis, and the difference between the two. Then you learned how the p-value is used to determine which hypothesis you accept and reject when compared to alpha. This led to type I and type II errors – what they are as well as how they interact with alpha. Finally, we went over what makes a good question for hypothesis testing and how to deal with a question that is less than ideal.

This has been a lot of theory in one chapter, but these are concepts that you must thoroughly understand, not only for the exam but also to perform any hypothesis testing as a data analyst. In the next chapter, we will cover some of the analyses you can use to perform a hypothesis test!

Practice questions

Let's try to practice the material in this chapter with a few example questions.

Questions

1. An e-commerce website runs an A/B study comparing two different website designs to see which generates more sales. What is the null hypothesis?

    A. There is a statistically significant difference between group A and group B

    B. Group A will be higher than group B

    C. Group B will be higher than group A

    D. Statistically, there is no significant difference between group A and group B

2. The A/B study from the previous question was finished. With an alpha of 0.05, the results were a p-value of 0.04. What do these results mean?

    A. You should accept the null hypothesis and reject the alternative hypothesis

    B. You should accept the alternative hypothesis and reject the null hypothesis

    C. You should accept the null hypothesis and the alternative hypothesis

    D. You should reject BOTH the null and alternative hypotheses

3. A small pet shop ran a study to see whether dog owners bought more pet food than other pet owners. After the study, they decided that there was no difference between the two groups, but they were mistaken – there actually is a difference. Dogs do eat more than most other pets. What type of error does this represent?

    A. Type I error

    B. Type II error

    C. Type III error

    D. Type IV error

4. You run a hypothesis test to compare the results of a new system to an old system. Using an alpha of 0.05, which of the following p-values would lead you to believe there is no difference between the new system and the old system?

    A. 0.03

    B. 0.008

    C. 0.05

    D. 0.4

5. If, in the scenario from the previous question, you find that there is a difference between the new system and the old system when there actually isn't. What type of error is this?

    A. Type I error

    B. Type II error

    C. Type III error

    D. Type IV error

Answers

Now, we will briefly go over the answers to the questions. If you got one wrong, make sure to review the topic in this chapter before continuing.

1. The answer is: There is no statistically significant difference between group A and group B

    Remember that the null hypothesis is that the two groups are the same and there is no statistically significant difference between them.

2. The answer is: You should accept the alternative hypothesis and reject the null hypothesis

    When your p-value is equal to or smaller than your alpha, you accept H_1 and reject H_0.

3. The answer is: Type II error

    In this scenario, the analyst accepted the null hypothesis and rejected the alternative hypothesis when they should not have. This describes a type II error.

4. The answer is: 0.4

    The value of 0.4 is the only number larger than 0.05. Because this p-value is larger than the alpha, we accept the null hypothesis, that there is no difference between the two groups, and reject the alternative hypothesis, that there is a difference.

5. The answer is: Type I error

    If you reword the question, you will find that you are falsely accepting the alternative hypothesis and rejecting the null hypothesis. This is a type I error.

10

Introduction to Inferential Statistics

In this chapter, we get into actual analyses—the fun part of data analysis. In this chapter, you will need to know what **t-test**, **chi-square**, **correlation**, and **simple linear regression** analyses are, as well as when to use them. We also included some hands-on practice for these analyses to help reinforce how and why each analysis is used. If you like, you are free to follow along and practice on your own, or you can simply read through and move on. Do what makes you feel confident in your understanding of the material.

In this chapter, we're going to cover the following main topics:

- Understanding t-tests
- Knowing chi-square
- Calculating correlations
- Understanding simple linear regression

Technical requirements

If you do decide to follow along, the data and code used can be found at the following URL:

https://github.com/PacktPublishing/CompTIA-Data-DAO-001-Certification-Guide

Understanding t-tests

In this section, we will cover the **Student's t-test**. This is a staple analysis, and some form of it is often used in hypothesis testing. Because of how useful it is, how often it comes up, and how relatively simple it is, a t-test is often the first inferential analysis data analysts learn. There are several different forms of this, but the main three are as follows:

- Independent t-test
- Dependent t-test
- One-sample t-test

These are pretty much what they sound like. An independent t-test, sometimes called an **unpaired t-test**, compares two groups that are independent of each other; the samples include different people or observations, gathered at different times, under different conditions. A dependent t-test, sometimes called a **paired t-test**, means that the two groups you are comparing are inherently related. If you collect data on a group of people at the start of the year and then collect data on the same group of people a year later, the two groups are dependent. Finally, a one-sample t-test takes one group and compares that entire group against a single value. For example, you might have the test scores of a single class and the mean of the scores of everyone who has ever taken that test. You are comparing a group to some outside value.

Technically, a t-test generates a t-score, which can then be used to determine statistical significance, but almost all data analytics tools will automatically give you the p-value as one of the results.

What you need to know about t-tests

For the exam, you will not be asked to perform a t-test, but you may be asked the definition of a t-test or be given a scenario and asked which analysis is most appropriate. You need to know that a t-test compares two groups that contain quantitative data. If you need to compare two columns of numbers, a t-test is your go-to analysis for this exam.

For example, if you want to compare the number of sales per week this year against the number of sales per week last year, you will run a dependent t-test. If you want to compare the number of sales per week this year against the number of sales per week this year for a competitor, you will run an independent t-test. If you want to compare the number of sales per week this year against the mean value for the industry, you will run a one-sample t-test.

T-test practice

Let's be clear that you will not need to run this analysis and you do not need to know any specific programming language for this exam. The hope is that a little extra practice will help cement this analysis in your mind so that you will be clear on how it differs from the other analyses.

T-test assumptions

First, for this example, we will be running a two-tailed independent t-test, so we need to know the requirements. The assumptions for an independent t-test are as follows:

- Independence
- Normality
- Homogeneity of variance
- $n \geq 30$
- The independent variable is categorical
- The dependent variable is numerical

The assumption of independence is simply that the two groups are independent of each other and that each observation is independent of the other.

The assumption of normality is that the data is normally distributed.

The assumption of homogeneity of variance is simply that the variance of both groups is around the same. If one variance is a little higher or lower than the other, it isn't a deal breaker, but if group A has a variance of 1 and group B has a variance of 10,000, it makes it really hard to compare them evenly.

This last one is not an actual assumption but rather, best practice. Technically, you can run a t-test with four observations in each group, or fewer. A t-test has no minimum sample size, but as a general rule of thumb, you get better results if you have at least 30 observations in each sample.

If you recall, an independent variable is what you are controlling or changing; in this case, the independent variable is what separates your two groups. Whether they are group A or group B derives your independent variable, and it doesn't matter what you call them.

The dependent variable needs to be a number since the t-test compares two groups of numbers.

All we need to start are two sets of normally distributed numbers that were collected separately, have similar variance levels, and have at least 30 observations each.

T-test code

For this example, we will be using Jupyter Notebooks to run Python. There are many ways to do this, using several different tools. Even if you want to use Python, there are several different packages that will allow you to do the same thing. The data we will be using is a sample from a pet database. Only four breeds of dogs and cats were selected for this sample: a random small dog breed (Yorkshire Terrier), a random large dog breed (English Mastiff), a random small cat breed, (Singapura), and a random large cat breed, (Maine Coon). The dataset includes Type, Breed, Weight (lb), Height (in), and Sex of the pets, with 200 observations.

The very first thing we do is gather everything we need. This involves importing all of the packages for the upcoming tests:

```
#Importing modules
import pandas as pd
import numpy as np
import statsmodels.api as sm
from scipy import stats
from sklearn.model_selection import train_test_split
```

To be clear, you are not going to need all of these packages for the t-test.

> **Important note**
>
> If you are using Jupyter Notebooks through the Anaconda distribution, you will already have all of these packages installed, so you only have to import them. Otherwise, if you have never used these packages, you will need to install them first, or importing them will fail. There are lots of guides online that go over how to install these things if you are having trouble.

It is a best practice to have all of your `import` statements together, so we have included every package we will need for this chapter.

Before we go any further, we need to pull in our data:

```
#Importing dataset
MyData = pd.read_csv("PetExample.csv")
MyData.head()
```

Okay, here, we are creating a variable called `MyData`, and using it to hold out the dataset. In the next line, we are using the `head` function to look at the first couple of rows. We can see the results in *Figure 10.1*:

	Type	Breed	Weight (lb)	Height (in)	Sex
0	Cat	Maine Coon	15	11	Male
1	Dog	English Mastiff	223	32	Female
2	Cat	Singapura	6	8	Female
3	Cat	Singapura	4	7	Female
4	Cat	Singapura	8	8	Male

Figure 10.1 – Preview of dataset

It is a best practice to preview the dataset when you first pull it in before you use it. This makes sure the data is what you think it is and you can double-check that all of the expected variables are present.

Okay! Now we have pulled in our required packages and data, let's jump right in. First things first—we need to prepare the data for this specific analysis:

```
#Preparing the data
SmallCats = MyData[(MyData["Breed"]=="Singapura")]
SmallDogs = MyData[(MyData["Breed"]=="Yorkshire Terrier")]
```

For this method of running a t-test, we first need to subset our data. Each group we are comparing needs to be stored in its own variable. Because we are using DataFrames from pandas, we can use the preceding logic to filter for all observations that are marked with a specific breed. The `SmallCats` variable now stores every entry where the breed is Singapura and `SmallDogs` holds every entry where the breed is Yorkshire Terrier. Now, we can compare these breeds directly. To double-check, if you run the `head` function on `SmallCats`, you will get a result similar to *Figure 10.2*:

	Type	Breed	Weight (lb)	Height (in)	Sex
2	Cat	Singapura	6	8	Female
3	Cat	Singapura	4	7	Female
4	Cat	Singapura	8	8	Male
9	Cat	Singapura	5	8	Female
13	Cat	Singapura	5	7	Female

Figure 10.2 – SmallCats preview

As you can see, we still have every column of information. If you recall, we are only comparing two numerical variables (one for each group), so we have to subset the data again. If you are comfortable, there are shortcuts to streamline this process, but until you get the hang of it, we will go ahead and define the specific variables we will be using:

```
#Isolating the test variables
SmallCatsWeight = SmallCats["Weight (lb)"]
SmallDogsWeight = SmallDogs["Weight (lb)"]
```

Here, we are isolating the specific variables we will be using for each group. In this case, we want to see whether there is a significant difference in weight between Yorkshire Terriers and Singapuras. Now, if you use the `head` function on `SmallCatsWeight`, you will see something similar to *Figure 10.3*:

```
2     6
3     4
4     8
9     5
13    5
Name: Weight (lb), dtype: int64
```

Figure 10.3 – SmallCatsWeight preview

You see two columns of numbers, but the row on the left is the index, or where these values fit into the dataset as a whole. This is isolated to just the weight of Singapuras. Everything is ready to plug into our t-test! Let's go:

```
#Running the analysis
tTestResults = stats.ttest_ind(SmallCatsWeight,SmallDogsWeight)
print(tTestResults)
```

In this case, we are creating a new variable to hold our analysis. `stats.ttest_ind()` is the function that runs our test. `SmallCatsWeight` and `SmallDogsWeight` are the variables we are feeding into the test. Finally, we are printing the results. If you were to run this code, you would get something like *Figure 10.4*:

```
Ttest_indResult(statistic=2.1530549587998804, pvalue=0.033930048809027155)
```

Figure 10.4 – T-test results

This is what we wanted to know—specifically, the p-value. Since we are using an alpha of 0.05, the p-value of 0.03 is significant. Here, we accept the alternative hypothesis—that there is a difference between them—and reject the null hypothesis—that there is no difference between them. There is a significant difference in the weight of Yorkshire Terriers and Singapuras. If you are curious, you can use the `describe` function on each test variable. The results for Singapuras will be as seen in *Figure 10.5*:

```
count    53.000000
mean      6.094340
std       1.362466
min       4.000000
25%       5.000000
50%       6.000000
75%       7.000000
max       8.000000
Name: Weight (lb), dtype: float64
```

Figure 10.5 – SmallCatsWeight.describe() results

You can do the same thing with the Yorkshire Terrier weight variable to get a result, as seen in *Figure 10.6*:

```
count    41.000000
mean      5.536585
std       1.074653
min       4.000000
25%       5.000000
50%       6.000000
75%       6.000000
max       7.000000
Name: Weight (lb), dtype: float64
```

Figure 10.6 – SmallDogsWeight.describe() results

The Singapuras' mean weight is 6.1 lb and the Yorkshire Terriers' mean weight is 5.5 lb. There is only a 0.6 lb difference between them, but we did find this difference to be statistically significant. Singapuras weigh more than Yorkshire Terriers on average.

If you want extra practice, you can try comparing the height of Yorkshire Terriers and Singapuras, the weight of cats and dogs, the height of Singapuras and Maine Coons, or any other combination you think is interesting. Play around with the data, have fun, and try to create a pleasant memory associated with t-tests.

This wraps up the basics of t-tests. Next, we will jump into another common analysis: chi-square!

Knowing chi-square

Here, we will go over everything you need to know about chi-square for the exam and go over a practical example. Chi-square is another useful tool that every analyst should have in their toolbox. As with a t-test, this technically produces its own value called a **chi-square statistic**, but, again, we can get a p-value out of it, and that is all we practically need. There are two popular kinds of chi-square tests, and you may run into either one on the exam, so make sure you understand both:

- Chi-square goodness of fit
- Chi-square test for independence

Unlike t-tests, these two types do very different things. There is actually a third type—a chi-square test for homogeneity—but it is rarely used, and CompTIA decided not to add it to the exam. Let's jump right in and get to what you need to know.

What you need to know about chi-square

A chi-square goodness of fit test is exactly what it sounds like: you are comparing a sample to a population to see whether that sample is a good representation. Want to know whether your sample truly reflects the data as seen in the population to make sure your analyses have any meaning whatsoever? This is your go-to. In the exam, any question that specifically asks to compare a sample to a population should make you think of this first.

A chi-square test for independence compares two categorical variables to see whether there is a relationship between them, or whether they are independent of each other. Here, the null hypothesis is that there is no relation between the two variables, and the alternative hypothesis is that there is a relation between the two variables. If you are asked which analysis would be most appropriate to compare two categorical variables, this should be your go-to.

Note—goodness of fit and test for independence are types of chi-square tests. There are several different versions of each of these types that may go by different names, but they will all clearly denote which type they fall into.

Chi-square practice

For this example, we will be using the same dataset with the same packages as we did for the t-tests before. You can pick up exactly where you left off, in the same code you used for the t-tests. Here, we will specifically be performing a chi-square test for independence (sometimes called **Pearson's chi-square**) to compare pet Type and Sex variables to see whether there is a relationship between them.

Chi-square assumptions

Okay, before we jump into assumptions, there is something you need to understand about a chi-square test for independence: it compares two categorical variables by analyzing a contingency table. Now, a contingency table is just another name for a frequency table that looks at more than one variable. We practiced making frequency tables earlier in this book in *Chapter 8, Common Techniques in Descriptive Statistics*. The main reason you need to know this is that some of the assumptions are based specifically on this contingency table. Here are the assumptions:

- Both variables are categorical
- Independence of observations
- Contingency cell exclusivity
- 80% of cells should have a value of at least 5
- $n \geq 50$

For chi-square, you are comparing two variables, and both of them must be categorical. The independence of observations is half of the assumption of independence that we covered with the t-tests. The variables themselves do not need to be independent, but each observation needs to be independent of every other.

Contingency cell exclusivity is simpler than it sounds. It just means that each observation is only counted once in the contingency table. You can't have observations that fall into more than one category and are counted multiple times. Either it is in one cell or another, but it can't be in both.

The next assumption is 80% of the cells on your contingency table should have a count of 5 or more. If you have a table with 10 cells, only 2 of those cells can have a value below 5. The general idea behind this is that if most of your cells have a really low count, it is difficult to get an accurate measurement.

Finally, $n \geq 50$. Chi-square is sensitive to sample size issues. It really helps if you have a decent number of values in most, if not all, of your cells. This leads us to a minimum sample size of 50.

> **Important note**
>
> Another common practice is to take the number of possible outcomes of one variable and multiply it by the number of possible outcomes from the second variable. This gives you the number of cells. Then, you multiply the number of cells by 5. For example, if you are comparing Color (Red, Blue, Yellow) to Shape (Circle, Square, Triangle), each has 3 possible outcomes. 3 X 3 = 9. There would be 9 cells in the contingency table. To get 5 observations in every cell you would need, at least, 9 X 5 = 45: 45 observations.

As long as you meet the other assumptions, the minimum sample size is flexible for this one. Pick a method that works for you and stick with it. Just remember that the larger your sample size, the more likely you are to have enough values in all of your cells, and the more likely you are to get accurate results.

Chi-square code

Let's jump right into the code. Since we covered all of the code to get the packages installed and the data imported when we discovered t-tests, we can save some time here by not repeating ourselves. Instead, we will create a contingency table that contains our two categorical variables: Type and Sex. With the pandas package, we can do this with a single line of code:

```
#Preparing the data
Contingency = pd.crosstab(MyData["Type"],MyData["Sex"])
print(Contingency)
```

Here, we are creating a variable called Contingency, then using the pd.crosstab function to fill it with a contingency table using our two variables of choice. Then, we print the results. The output should look something like *Figure 10.7*:

```
Sex    Female  Male
Type
Cat        47    64
Dog        50    38
```

Figure 10.7 – Contingency table

This should look familiar, as a rather simple frequency table. We see in this sample that there are more female dogs than male dogs and more male cats than female cats, but is this just chance, or is there a relation between these two variables? The only way to know for sure is to run our analysis:

```
#Running the analysis
stat,p,dof,expected = stats.chi2_contingency(Contingency)
print(p)
```

You might notice that this looks slightly different. This test actually uses the contingency table to create four different output variables: stat, p, dog, and expected. For now, the only one we care about is p, which holds the p-value of this analysis. Let's look at the results of our chi-square analysis in *Figure 10.8*:

0.0592618016299914686

Figure 10.8 – Chi-square p-value

This p-value is larger than our alpha of 0.05, if only by a little. This means that we accept the null hypothesis and reject the alternative hypothesis. For a chi-square, that means that we are saying that, despite how the numbers may look in the table, there is not a statistically significant relationship between the Type and Sex variables.

That's it for chi-square, so let's move on to another common analysis: correlation.

Calculating correlations

There are many different versions of correlation analysis, but they all do the same basic thing and you will not be required to distinguish between them. Before we jump into talking about correlation analysis, let's take a moment to discuss the concept of correlation in a broader sense.

Correlation

A correlation is simply a relationship between two variables. If there is a relationship, it can be positive or negative. Earlier, in *Chapter 6, Types of Analytics*, we discussed using scatter plots to get a general idea about these relationships. Let's use them as illustrations now. In *Figure 10.9*, we see what a positive correlation looks like:

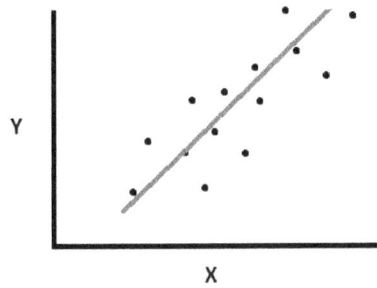

Figure 10.9 – Positive correlation

A positive correlation means that as one variable increases, so does the other. This means that the dots form a loose line from the lower left to the upper right. For example, there is a positive correlation between the number of customers and the number of sales, because as you get more customers, there are more people to buy things, so you make more sales. Now, let's look at a negative correlation in *Figure 10.10*:

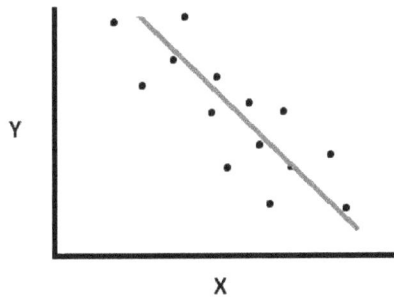

Figure 10.10 – Negative correlation

In a negative correlation, as one variable increases, the other variable decreases. This creates a rough line from the upper left-hand corner to the lower right. For example, cats and mice share a negative correlation. As the number of cats in a house increases, the number of mice in the house should decrease… because they get eaten. Let's look at what happens when there is no correlation, as in *Figure 10.11*:

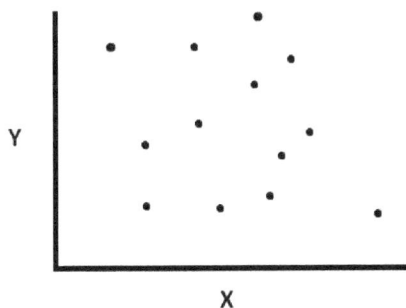

Figure 10.11 – No correlation

When there is no correlation, it means that what happens to one variable has no impact on the other. This means the scatter plot looks like a truly random assortment and is not in the shape of any line.

However, the most important lesson is this:

CORRELATION IS NOT CAUSATION

This is not only important to know for the exam, but for life working as a data analyst. Just because you know there is a correlation between two variables does not mean you understand *why* they are correlated. You cannot say for sure whether changes in variable A are causing changes in variable B, whether changes in variable B are causing changes in variable A, or whether changes in both variables are being caused by some mysterious variable C. The fact is that you cannot tell why these relationships exist from a correlation analysis, only that they do.

Okay, what if you aren't sure from glancing at a scatter plot how correlated two variables are? For times such as these, we want a hard number to tell us exactly how strong the relationship between two variables is. That number is called the correlation coefficient, and we find it through a correlation analysis.

The correlation coefficient ranges from -1 to 1. Let's see this in *Figure 10.12*:

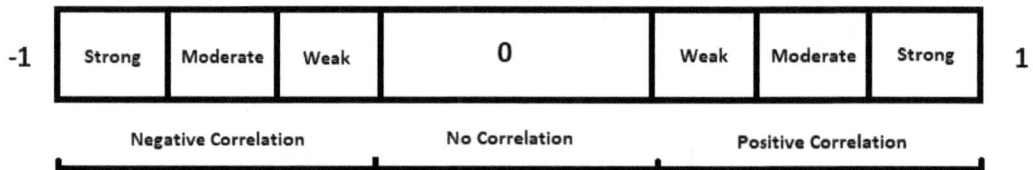

-1	Strong	Moderate	Weak	0	Weak	Moderate	Strong	1

Negative Correlation   No Correlation   Positive Correlation

Figure 10.12 – Correlation coefficient

The specific cutoff scores for each category are a matter of some debate, but you should understand that the closer the correlation coefficient is to 1 or -1, the stronger the relationship. Conversely, the closer the coefficient is to 0, the weaker the correlation is until there is no correlation at all.

What you need to know about correlation

While chi-square analysis tests to see whether or not there is a relationship between two categorical variables, correlation analysis tests to see how strong a relationship is between two numerical variables. Note that there is a special correlation analysis called **Spearman's correlation** that compares a numerical variable to the levels of an ordinal variable. However, this is not the most common type of analysis and is unlikely to come up in the exam. Still, if you are asked a question as to which analysis is most appropriate to see whether there is a relationship between a numerical variable and anything else, correlation analysis should be your first thought.

Also, make sure you can identify negative correlations, no correlations, and positive correlations on sight, whether they give you a correlation coefficient or a scatter plot.

Correlation practice

For our example, we will be running a Pearson's correlation analysis to see whether there is a relationship between the weight of the pets and their size. Pearson's is the most common type of correlation analysis and looks at two numerical variables: in this case, `Weight` and `Height`.

Correlation assumptions

Looking specifically at the assumptions for a Pearson's correlation, we will see some assumptions we have run into before and some we have not:

- Level of measurement
- Linearity
- Normality
- Related pairs
- Lack of outliers
- n ≥ 30
- Two continuous variables

The level of measurement can get needlessly complicated very quickly. There are four levels that give increasingly more meaning to the values of numbers. You can make arguments until you are blue in the face, but if you have two continuous variables (which you should have for this analysis anyway), then you meet this requirement. In our case, height and weight are both physical measurements, where the distance between one number and another has value, which technically means their level of measurement is Interval, which is the second highest in the order. Long story short, have two continuous variables and you knock out two assumptions in one go.

Linearity is a little more straightforward. If you do a scatter plot and draw a line through the dots, it should be a straight line. If there is a defined shape to the dots, but it is a wave or a curve instead of a straight line, then you don't meet this one. That said, if it is just a random ball of dots, then go ahead and try!

We have talked about normality before, so we can skip that one for now.

Related pairs mean that for every data point in one variable, there should be a matching one in the other. In this case, every pet has a weight measurement and a height measurement, so there is a specific pair of values for every observation.

A lack of outliers is exactly what it sounds like. You should not have any outliers in your dataset. This is a best practice for any dataset, but a single outlier can ruin the results of this analysis.

Generally speaking, a larger sample size is better, up to a point, but the accuracy of correlation analysis benefits from this more than most analyses. A minimum sample size of 30 ensures a minimum level of accuracy.

Correlation code

The actual code for this one is straightforward. Let's start by defining our variables:

```
#Preparing the data
PetHeight = MyData["Height (in)"]
PetWeight = MyData["Weight (lb)"]
```

This is just setting up our `PetHeight` and `PetWeight` variables and filling them with the appropriate data from our dataset. Next, we run the analysis!

```
#Running the analysis
CorrelationResults = stats.pearsonr(PetHeight,PetWeight)
print(CorrelationResults)
```

Again, we are creating a variable and then filing it with the `stats.pearsonr` function using the variables we prepared ahead of time. Let's look at the results in *Figure 10.13*:

```
(0.9592991777202481, 3.8891158549349756e-110)
```

Figure 10.13 – Correlation results

This returns two values: the correlation coefficient and the p-value, respectively. A correlation coefficient of 0.96 would definitely count as a strong positive correlation. The p-value is so ridiculously tiny that we must accept the alternative hypothesis and reject the null hypothesis. There is a relationship between these two variables—a strong one.

This wraps up correlations, so let's move on to simple linear regression.

Understanding simple linear regression

Regression is an entire category of statistical analysis, which includes dozens of types are variations. Even within those types, different people will use the same analysis in different ways. However, the first kind that most people learn is simple linear regression. It's called **simple** because it only has one independent variable, and **linear** because it draws a straight line.

What you need to know about simple linear regression

Simple linear regression is all about prediction. You are testing to see whether one variable is a good predictor of the other one. If it is, you can use this model to actually make predictions. First, let's talk about the chart:

- The x axis is the independent variable (predictor variable)

- The y axis is the dependent variable (criterion variable)

Along the bottom is the x axis, and this is where you put your independent variable. In regression analysis, this is called the predictor variable, because the hope is that you can use this variable to predict what the other one will be. Along the side of the chart, you have the y axis, and this is where you have your dependent variable, called the criterion variable in regression analysis, or the variable you are trying to predict. We can see what this looks like in *Figure 10.14*:

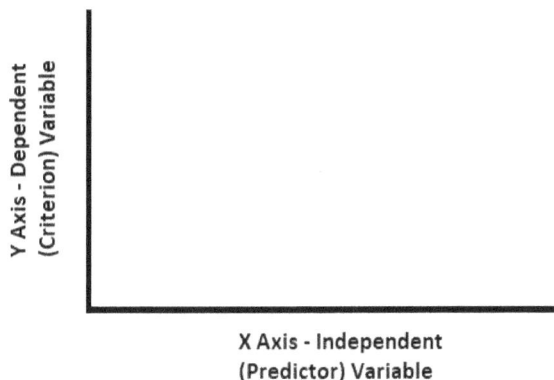

Figure 10.14 – Linear regression variable placement

When you run the analysis, you can get a p-value. In this case, you are testing whether the independent variable is a predictor of the dependent variable. The null hypothesis is that the independent variable is *NOT* a predictor of the dependent variable, and the alternative hypothesis is that the independent variable *IS* a predictor of the dependent variable.

While calculating this process, a straight line is drawn through the data. This calculated line is your model. You will usually also get an R^2, called the coefficient of determination. This number tells you how much of the variance in your dependent variable is explained by your independent variable. In other words, it gives you a guide to how well the model fits with your data. R^2 ranges from 0 to 1 and translates to a percentage. For example, if your R^2 is 0.50, that means that 50% of the data points fall within the regression line.

If your p-value is low enough and your R^2 is high enough, you can give new values for your independent variable to the model and get the corresponding dependent value. Sound a little squirrely? Well, then, let's look at an example.

You collect, feed, and train messenger squirrels. In addition to a balanced diet, you feed the squirrels nuts as treats. You feed each squirrel a different number of nuts, then weigh them. You want to see whether the number of nuts you feed them impacts how much they weigh. Let's look at your findings on a scatter plot in *Figure 10.15*:

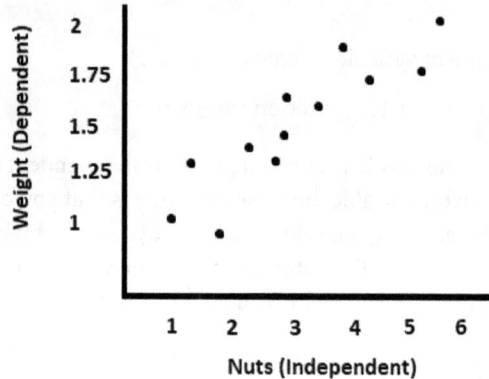

Figure 10.15 – Nuts and weight scatter plot

This seems pretty straightforward, so we run a simple linear regression on our data. The p-value is 0.043, so with an alpha of 0.05, we are comfortable saying that the number of nuts you feed the squirrels can predict their weight. This gives us a line through the data, as seen in *Figure 10.16*:

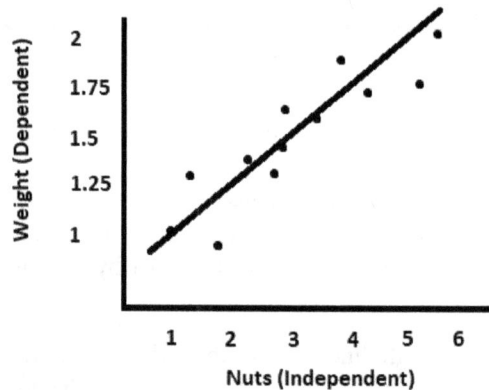

Figure 10.16 – Nuts and weight linear regression line

A lot of the time when a data analytics tool does a linear trend line or a best-fit line, it is actually calculating that line with simple linear regression. So what is the point of this line? It means that we can predict how much one of the messenger squirrels will weigh based on how many nuts you feed it every day. Let's see what happens when we feed a squirrel just one nut every day. This can be seen in *Figure 10.17*:

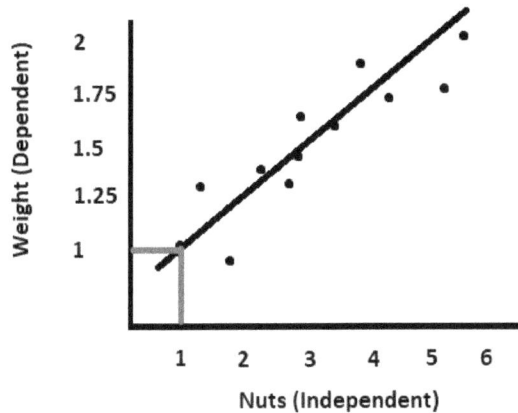

Figure 10.17 – Prediction of one nut

As we can see, feeding a squirrel only one nut every day will lead to it weighing only 1 lb. You are not, however, limited to values you have recorded before. You can predict future values. What if we were to feed a squirrel seven nuts a day? This is depicted in *Figure 10.18*:

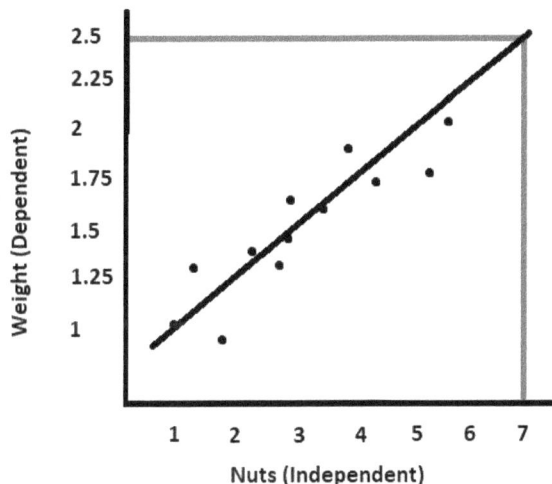

Figure 10.18 – Prediction of seven nuts

As we see in *Figure 10.18*, feeding a squirrel 7 nuts a day, on top of its normal diet, creates a very round squirrel that weighs 2.5 lb. Still, you can see how this ability to use one variable to predict another variable could come in handy.

If in the exam you are ever asked which test will tell you that one variable can predict another, or you are asked which analysis is best for predicting one variable using another, simple linear regression should be your answer. If the question has the word "predict" in it at all, you should start looking for signs that it is simple linear regression.

Simple linear regression practice

Here, we will see whether we can predict a pet's height by using its weight with simple linear regression. That means `Weight` is our independent variable and `Height` is our dependent variable. Before we can go into how to do this, we need to look at the requirements.

Simple linear regression assumptions

We have seen most of these assumptions before, but there is one that will throw us for a bit of a loop. The assumptions are as follows:

- Linearity

- Normality

- Independence

- Homoscedasticity

- The dependent variable is numeric

- The independent variable is numeric

- n ≥ 100

We have already talked about assumptions of linearity, normality, and independence, so we can skip those for now. Let's go over the concept of homoscedasticity. Technically, the definition of homoscedasticity is something about the variance in the residual values remaining constant for different levels of the independent variable. Let's break this into something useful. Residuals are just the distance every point is away from the line, so all this is saying is that the dots are roughly the same distance from the line all the way down its length. Let's see what this looks like in *Figure 10.19*:

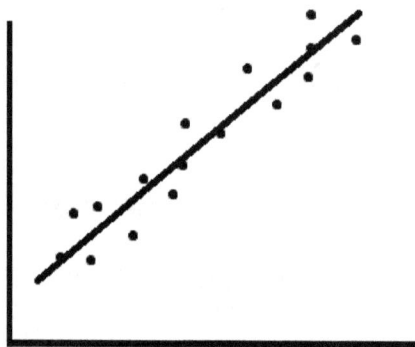

Figure 10.19 – Homoscedasticity

All of the points are roughly a similar distance from the line all the way down its length. This even spread of points is a good example of homoscedasticity. Let's take a moment to look at an example in *Figure 10.20* of something that would fail the homoscedasticity assumptions:

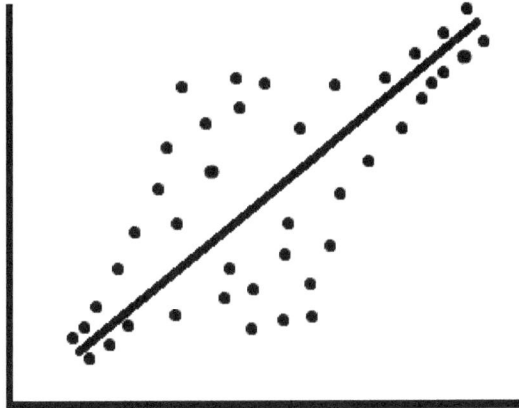

Figure 10.20 – Not homoscedasticity

In the preceding example, it starts out with the points all really close to the line, then the dots really spread out, then come back near the line. Even if this would pass the assumption of linearity, it would not pass the assumption of homoscedasticity.

Again, we are creating a scatter plot and drawing a line through it, so the independent and dependent variables both must be numbers. When it comes to the sample size, you can technically get away with 10 observations per independent variable. In simple linear regression, there is only 1 independent variable, so the minimum sample size is 10. That said, common practice is at least 100 observations for best results.

Simple linear regression code

Let's jump right in with preparing the data for simple linear regression:

```
#Preparing the data
Y = MyData["Height (in)"]
X = MyData["Weight (lb)"]
XConstant = sm.add_constant(X)
```

This is pretty straightforward. First, we define our X and Y variables, by filling them with Weight (lb) and Height (in), respectively. Now, for this package, we need to define a constant. We do this by plugging our X variable in the sm.add_constant() function and saving it in a new variable, XConstant.

Now, it's time to actually run the analysis!

```
#Running the analysis
SimpleLinearRegressionResults = sm.OLS(Y,XConstant).fit()
print(SimpleLinearRegressionResults.summary())
```

Okay, so we are doing a few things here. First, we are creating a `SimpleLinearRegressionResults` variable to hold the results.

> **Important note**
>
> Variable names should be short, simple, and clear. For the examples in the book, a lot of the variables have very explicit names that tell you exactly what they hold, such as `SimpleLinearRegressionResults`. This is to make the example as clear as possible. When you are working with your own code, you are encouraged to use something shorter, as long as it is clear what it is. For example, you could try `SLRMod` or `LinReg`. Just pick a naming convention and stick with it.

The `sm.OLS()` function creates a model using `Y` and `XConstant`, and the `.fit()` function applies it to our data. Just like that, the model is ready to go. That wasn't so bad, right? Now, this model has a few built-in functions. Here, we use the `.summary()` function to give us the results of our model. We can see what this looks like in *Figure 10.21*:

```
                            OLS Regression Results
==============================================================================
Dep. Variable:            Height (in)   R-squared:                       0.920
Model:                            OLS   Adj. R-squared:                  0.920
Method:                 Least Squares   F-statistic:                     2273.
Date:                Mon, 22 Aug 2022   Prob (F-statistic):           3.89e-110
Time:                        16:02:24   Log-Likelihood:                 -487.92
No. Observations:                 199   AIC:                             979.8
Df Residuals:                     197   BIC:                             986.4
Df Model:                           1
Covariance Type:            nonrobust
==============================================================================
                 coef    std err          t      P>|t|      [0.025      0.975]
------------------------------------------------------------------------------
const          8.5104      0.241     35.272      0.000       8.035       8.986
Weight (lb)    0.1220      0.003     47.680      0.000       0.117       0.127
==============================================================================
Omnibus:                       13.770   Durbin-Watson:                   2.314
Prob(Omnibus):                  0.001   Jarque-Bera (JB):               12.161
Skew:                           0.530   Prob(JB):                      0.00229
Kurtosis:                       2.413   Cond. No.                         114.
==============================================================================

Notes:
[1] Standard Errors assume that the covariance matrix of the errors is correctl
y specified.
```

Figure 10.21 – Simple linear regression results summary

This is a lot of data, but a data analyst is not a statistician, so we don't care about most of it. In the upper-right corner, we see R-squared, which is just another way of saying R^2. Our R^2 here is 0.92, which means 92% of the variance in height can be explained by weight. Next, we will find the p-value. Here, it is listed as P>|t|, but it is still just a p-value. Also, it is rounded to 0.000, so we can safely say that our p-value is pretty small. Here, we would accept the alternative hypothesis and reject the null hypothesis; you can predict, roughly, a pet's height based on how much it weighs.

Summary

We covered a lot of information in this chapter. First, we talked about t-tests and how they are used to see whether there is a significant difference between two groups that use numeric variables. Then, we ran right into chi-square tests, and how they have two main types: chi-square goodness of fit and chi-square test for independence. A chi-square goodness of fit test sees whether a sample is a good representation of a population, while a chi-square test for independence compares two categorical variables to see whether there is a relationship between them. Next, we talked about correlation and how it can be used to see whether two numeric variables are related and how strongly they are related. Finally, we talked about simple linear regression and how it can be used to see whether one numeric variable can predict another. This wraps up everything you need to know about analyzing data for the exam.

In the next chapter, we will move on to reporting data!

Practice questions

Let's try to practice the material in this chapter with a few example questions.

Questions

1. *Simon's Grocery Store* collects data on the number of sales of every vegetable it sells every day. If you wanted to see whether there is a statistically significant difference between the number of sales of pumpkins and the number of sales of summer squash, which analysis would be most appropriate?

    A. T-test

    B. Chi-square

    C. Correlation

    D. Simple linear regression

2. Now, *Simon's Grocery Store* groups cereal brands into Great Sellers and Poor Sellers based on sales performance in the past. It also notes the color of the box for each brand. The grocery store would like to know whether there is a relationship between the color of the box and whether the brand is a great seller or a poor seller. Which analysis would be most appropriate?

    A. T-test

    B. Chi-square

    C. Correlation

    D. Simple linear regression

3. *Simon's Grocery Store* collects data on every product it sells. Now, it is wondering whether there is a relationship between the number of turkeys sold and the number of packages of premade stuffing sold. Which analysis would be most appropriate?

    A. T-test

    B. Chi-square

    C. Correlation

    D. Simple linear regression

4. No matter the results of the previous question, *Simon's Grocery Store* wants to know whether it can predict the number of packages of premade stuffing it will sell based on the number of turkeys it sells. Which analysis would be most appropriate?

    A. T-test

    B. Chi-square

    C. Correlation

    D. Simple linear regression

Answers

Now, we will briefly go over the answers to the questions. If you got one wrong, make sure to review the topic in this chapter before continuing:

1. The answer is: *T-test*

    Here, you are comparing two numeric variables to see whether there is a difference between them, and that makes this a job for a t-test.

2. The answer is: *Chi-square*

    In this question, we are comparing two categorical variables to see whether there is a relationship between them. This is a good time to use a chi-square test for independence.

3. The answer is: *Correlation*

    For this question, we are seeing whether there is a relationship between two numeric variables. That means we are using correlation to see whether there is a relationship and how strong it is.

4. The answer is: *Simple linear regression*

    Here, we are seeing whether we can predict one numeric variable using another numeric variable. That means the first thing we should think about is simple linear regression.

Part 3: Reporting Data

This part guides you on the best way to report your results and do so in a way that communicates effectively and won't risk leading to any legal issues.

This part covers the following chapters:

- *Chapter 11, Types of Reports*
- *Chapter 12, Reporting PRocess*
- *Chapter 13, Common Visualizations*
- *Chapter 14, Data Governance*
- *Chapter 15, Data Quality and Management*

11
Types of Reports

In this chapter, we will go over the general categories of reports that will be covered in the exam. The average data analyst will be creating some—if not all—of the report types mentioned here in the course of their career. Some fields may use more of some types than others, but it is generally a good idea to know which types of reports might be expected of you. That said, your main focus is passing the exam, and we will tell you everything you need to know about static and dynamic reports, ad hoc and research reports, self-service reports, and recurring reports. To wrap up this chapter, we will talk about some data analytics tools you will need to know about for the exam.

In this chapter, we're going to cover the following main topics:

- Distinguishing between static and dynamic reports
- Understanding ad hoc and research reports
- Knowing about self-service reports
- Understanding recurring reports
- Knowing important analytical tools

Distinguishing between static and dynamic reports

In this section, we will briefly talk about static and dynamic reports. This designation is a little different than the other report types you will learn about in this chapter. To be clear, all reports will either be static (sometimes called **point-in-time**) or dynamic (sometimes called **real-time**). Any of the other types of reports (ad hoc, self-service, recurring, or research) will *ALSO* be either static or dynamic. The reason the report is static or dynamic has nothing to do with the contents of the report but, rather, the relationship between the report and the original data.

Point-in-time reports

Throughout most of history, static reporting was the only kind of reporting. The idea is relatively simple: if you collect data, stop, analyze data, and then report it, you have created a static report. The moment you stop collecting data, it will become static. Effectively, this kind of report is like a picture that gives the status of whatever you are collecting data on at that moment. If you were to collect data again a year later, you would have another snapshot of the data at another fixed point in time. You could then compare these two fixed points and figure out the differences, but you are still working with a series of fixed points. Even if you are continually gathering data, you create fixed points when you analyze the data and write something meaningful about it. Even if you write the most up-to-date, in-depth report you possibly can, next year, this report will still be last year's report. If you run it again, to keep it up to date, you are generating a new report.

Let's look at the pros and cons of static reporting.

Pros:

- Can use more in-depth and complicated analyses
- Can answer more complex questions
- Can pre-empt follow-up questions
- Can be generated faster

Cons:

- Data is a fixed point and not up to date
- Often needs to be generated multiple times

Static reports are powerful and give you more options to adapt to specific conditions. This means that they *can* be more complicated, and you need to take time to make sure you are communicating effectively.

Real-time reports

Dynamic reports are exactly what you think they are: reports with a live connection to the data. That means that they update automatically whenever there is a change in the data. Instead of a picture of the data, as with static reporting, dynamic reporting is like a live stream of the data. This is usually done through a dashboard or a web service. If you can imagine a web page that tells you the total number of times you click the button on it, and the moment you click the button, the page shows a new number, this is dynamic reporting. As soon as the data—the number of clicks—changes, it is reflected in the report.

Let's look at the pros and cons of dynamic reporting.

Pros:

- Data is always as up to date as possible
- Often saves time in the long run

Cons:

- Limited to relatively simple analyses
- Limited to answering relatively simple questions
- Can take a long time to set up

Dynamic data is a great tool; it doesn't have as much analytical power, but it is simple, requires relatively little maintenance once it is set up, and its greatest strength is that it always shows the most up-to-date information.

Static versus dynamic reports

There are a lot of people, usually business analysts or those who have to deal with data but are not data professionals, who think static reports are a thing of the past and should be phased out in favor of dynamic reports. However, this is short-sighted and will never happen. The truth of the matter is these two report types fill different roles.

Dynamic reporting shines in situations where having the most recent data is more important than having detailed analyses, or where the same simple questions get asked over and over again. Imagine you work with a stockbroker, and they can gain or lose millions of dollars depending on how the stocks change in a matter of minutes. Every second counts, and having the most recent stock data is crucial. In this case, it is irresponsible to use anything but dynamic reporting.

Static reporting has a few different roles. Often, if the question is never going to be asked again, it is pointless to spend time setting up a dynamic report. It is faster and easier to just answer the question and be done. For example, maybe your boss wants to know whether buying a new sanitizing product for the office will reduce the number of sick days employees take or whether it is just a meaningless expense. This is a one-time comparison. Either it is worth it, or it isn't. There is no need to run the same study every 3 months. The other place static reporting shines is when you have to answer a big, complicated question with several big, complicated analyses. Often, you take the results of one question, gain insight, and then ask another question. This process requires a more hands-on approach, and static reporting is really the best answer.

Let's not ignore the fact that too much automation will put data specialists out of a job. Your future career depends on your ability to give more meaningful insight into the data than a phone application. Fight on! Don't let the robots win! Next, let's talk about ad hoc and research reports.

Understanding ad hoc and research reports

Ad hoc and research reports are similar in some ways and opposite in others. They both answer questions directly, but one is short and sweet while the other is very thorough. Let's just jump right in, and you will see.

Ad hoc reports

Ad hoc reporting is an important part of a data analyst's job. Simply put, ad hoc reports are short, quick reports that answer a simple question. How often you will perform ad hoc reports depends heavily on the position and why it was created. Some positions will spend most of their time on ad hoc reports, while other positions will focus on other types of reports, such as recurring reports, and will hardly ever see an ad hoc report request. For example, healthcare analysts will often get questions from all over a hospital or set of hospitals about one thing or another. On the other hand, a business analyst may spend most of their time on recurring reports and seldom get a direct question. Even within these stereotypes, there will be companies or even departments that change what they require of a data analyst wildly.

Ad hoc reports are almost always static. There is no need to have the most recent information, and it will not be asked again. These will often have a short turnaround time as well, so taking the time to set up a dynamic report is usually counter-productive.

More often than not, there is someone in the company who has to make a decision. Instead of blindly guessing, they ask you a question so that they can make an informed decision. That decision is usually put on hold until you get back to them. The more pressing the decision is, the more pressure on you to get them an answer quickly. Sometimes, this could be something as pointless as asking whether the brand of coffee in the break room makes a bigger difference than the flavor of the coffee. Trying to figure out how to answer weird questions comes with the job, I'm afraid. Other times, you may get questions that will directly impact the company or the product. A toy manufacturer may want to revamp its toy blocks and needs to know whether it should add more colors or more shapes: which stimulates children more?

Ad hoc reports are all about answering relatively simple questions quickly and directly. These reports do *not* include a 50-page essay that debates the pros and cons and suggests a follow-up study. These reports are usually quick (often only a page) and give results and sometimes a brief description of what they mean: the brand is more important than flavor, and the color is more important than shape. They don't need a dissertation explaining *why* the answer is what it is—they just need an answer so that they can make a decision and move on. These reports are quick and to the point.

Research reports

Research reports, sometimes called tactical reports, are the big, bad reports. These are static, without question. A research report is a large, in-depth report that answers large, complicated business questions using every advanced statistical analysis you can possibly cram in there. When we talked about ad hoc reports, we mentioned they were not supposed to be dissertations; well, research reports are.

Often, these reports are supposed to answer big questions that will impact the direction of the company. Should we merge with Company B? Should we buy Company C? Should we sell our company to Company A? Should we pivot into a different field? Should we change our flagship product? Do we need to change our target audience? All of these are big questions, and reports will often try to address them by breaking them down into several smaller questions, answering each in turn in an attempt to address the big question as a whole.

Needless to say, these do not happen routinely, and they can take a lot of time and effort to generate. Sometimes, you will need to gather additional data or run an entire study to be able to make a research report. This also means that, for the company, research reports are often expensive, which is another reason they may not be very common.

Keep in mind that research reports, as with any other reports, depend heavily on the company. It is possible that you will find yourself in a position where you are expected to write short five-page research reports every week. There is no way to know for sure, but for the purpose of the exam, know that research reports are longer, more in-depth reports that answer more complicated questions.

This wraps up ad hoc and research reports, so let's move on to self-service reports.

Knowing about self-service reports

Self-service reports are becoming more and more common as companies continue to develop software specifically for this purpose. There are now several different kinds of self-service reports, but the most common by far is the dashboard report. A dashboard uses software, an application, or a web service to give limited access to specific metrics, data, or visualizations.

Why are these increasing in popularity? Because after they are set up by a data analyst, anyone you give access can use the application to answer their own questions. This is especially important when there are positions in the company that require non-data specialists to interact with data. Consider a mortgage company: a mortgage agent may not be a trained data professional, but they need to be able to access things such as the current interest rate quickly and easily when working with clients. This way, they don't need to put the client on hold, call you up, and ask you to give them the numbers they need right away while the client waits. They can just look it up themselves. Any data that commonly needs to be used, especially if several different people need to use it, can be made accessible through self-service reports. You may even have an application on your phone that you sign in to to receive up-to-date data on a particular topic.

Let's take a moment to talk about how data is connected to a self-service report. A lot of dynamic reports are self-service reports. It is relatively easy to connect a dashboard directly to your data source to stream live data. However, not all self-service reports are dynamic. You can absolutely make a dashboard with data you pulled from a query. In this way, whatever you are viewing on the dashboard will not change until you manually change the data. This is even a relatively common practice. You can manually refresh the data in a dashboard or set some stimulus or schedule to refresh the data. However, this is still considered a static report; it's just a static report that you keep updating. There is no constant connection between the dashboard and the data, which is required to be considered dynamic.

Self-service reports are almost always simple analyses and visualizations based on metrics. Don't get me wrong—some people take dashboards to another level. If you are interested in what these might look like, visit `public.tableau.com/app/discover/viz-of-the-day`. Tableau publishes and shows off some of the most spectacular dashboards and visualizations created by data analysts just like you on its website. It is, quite literally, an art form. That said, there is no need to compete with them—the majority of functional self-service reports are simple and clean and report the metrics that the audience needs to do their job. That's all you need to know about self-service, so let's move on to recurring reports.

Understanding recurring reports

Onward to recurring reports. These, as you might have guessed, are reports that occur at regular intervals. For some data analysts, this is the majority of what they do. Depending on the industry, company, department, or even team, there are all kinds of reports that will need to be given at regular intervals. Some of these are more in depth and can include more advanced analyses, depending on the topic.

Recurring reports are usually static. This is for a couple of reasons. These reports can contain more complicated analyses, and they may require specialized insight or interpretations, but the most important reason is that a lot of these reports are given at meetings so that they can be discussed immediately to make decisions. Yey, some recurring reports may be replaced with self-service reports, but these may not be things that people will need to interact with on a regular basis, so self-service reports will often be ignored. A meeting with a recurring report forces people to pay attention to the data, if only for a few minutes.

This exam breaks this down into three common categories of recurring reports:

- Compliance reports
- Risk and regulatory reports
- Operational reports (KPI reports)

For the exam, you will need to know what these are individually, but also that they are all considered recurring reports.

Compliance reports

The first kind of recurring report you need to know about is a compliance report. This gets a little confusing because compliance reports are sometimes called **regulatory reports**. The general idea behind compliance reports is to make sure you are compliant with a set of regulations. This is where you make sure you are meeting every requirement imposed upon you. You can think of this as asking yourself whether your company is coloring within the lines.

Regulations come from all different levels. They can be the following:

- Internal
- Third-party
- Industrial
- Governmental

Internal regulations are some set of requirements that the company sets up itself or rules and goals that *must* be followed. A third-party regulatory service is usually optional. If you want this other company to certify you and give you their stamp of approval, you need to meet their requirements. Industry regulations are general rules and requirements that must be met when working within a specific industry. For example, pharmaceutical data has very specific ways in which it must be treated that are not required in any other field, because mistakes can lead to accidentally poisoning thousands to millions of people. Finally, government regulations are probably the strictest set of rules. Not being in compliance with a set of regulations, no matter the level, usually has harsh consequences. It is in everyone's best interest to make sure that all requirements are met, and it's often your job to check and report whether or not they are.

Risk and regulatory reports

Risk and regulatory reports are not the same as compliance reports, but they are similar. These reports are all about risks; often, they are identifying the risk of regulations changing, but they also cover the impact of those risks or regulation changes. What are the chances of this happening, and if it does happen, how bad is the damage? If this new bill passes, how will it impact the regulations of the company? If the regulation changes from this to this, how much will it cost them to meet the new regulations? Sometimes, this will even cover sketchy conversations, such as is it cheaper to be out of compliance and accept the fines than it is to change to meet them?

Operational reports (KPI reports)

Next, let's talk about operational reports. These are, by far, the most common form of recurring reports. Compliance and risk and regulatory reports are often only once a year, but operational reports are more likely to be quarterly. Operational reports are only there to report the current state of normal operations. This is just a quick check to see how things are going in general. More often than not, these reports focus on KPIs. Again, which KPIs are reported depends heavily on what you are doing, but it is common for companies to check regularly to see whether they are gaining money or losing money.

This is everything you need to know about recurring reports for the exam. It's time to move on and start talking about some analytical tools you might use to actually get these reports.

Knowing important analytical tools

Okay, this one is brute-force memorization. You do not need to know how to use any particular data analytics tool or software for the exam, but CompTIA+ has come up with a list of names you need to know. There may be tools on the list that aren't very popular, and there may be popular tools that aren't mentioned on the list, but this is a list of the tools that will be in the exam. To be clear, this is all very high level—you just need to memorize the names and the vague purpose. The full list is as follows:

- **Structured Query Language (SQL)**
- Python
- Microsoft Excel
- R
- RapidMiner
- IBM Cognos
- IBM **Statistical Package for Social Sciences (SPSS)** Modeler
- IBM SPSS
- **Statistical Analysis System (SAS)**
- Tableau
- Power BI
- Qlik
- MicroStrategy
- BusinessObjects
- Apex
- Dataroma
- Domo
- **Amazon Web Services (AWS)** QuickSight
- Stata
- Minitab

This is quite a list, and you probably have not heard of all of these. To be clear, no one uses all of these, and a few of them are definitely not as popular as they used to be. Also, they have a wide range of uses, and some of these actually represent multiple things. There is no easy trick to learning these, but people often benefit by breaking the list into chunks. Let's go ahead and break these down into some general categories by their purpose.

Query tools

This one should be pretty easy to remember. Here it is:

- SQL

That's it. The only query language on this list is SQL, but let's take a moment with this. SQL is a programming language that is designed to work with a **relational database management system (RDBMS)**. The original program was developed by IBM, but since then, there have been literally dozens of different languages based on SQL, with the most popular being PostgreSQL, SQLite, MySQL, and SQL Server. Luckily, these all have very similar syntax. If you know one, you can probably figure out the others.

It is worth mentioning that there are several query languages that are not based on SQL, and these are gradually gaining in popularity, but at the moment, they are not something that the exam thinks you need to know.

If you are interested, information on SQL is easy to find.

Spreadsheet tools

Spreadsheet tools are for your basic flat files. The one you need to know for sure is this:

- Microsoft Excel

Again, there are several spreadsheet software tools that exist, such as Google Sheets, LibreOffice, and Zoho Sheet. Microsoft Excel was chosen because it is the most common, and it is the one that companies are most likely to request a data analyst to use. You just need to know that Microsoft Excel is spreadsheet software.

Programming language tools

This is where it starts to get a little more interesting. These are tools that involve writing code:

- Python
- R
- SAS
- Stata

Python is the only program on this list that is a general-purpose programming language. R, SAS, and Stata are all programming languages developed specifically for statistical analyses. These are not quite as similar as query languages or spreadsheet software. If you learn one thoroughly, you will have less trouble learning the others; a lot of the basic concepts are the same, but it will take more work to learn more than one of these.

Visualization tools

Of course, these tools can do all kinds of things, but their main purpose is to quickly and easily create visualizations with easy point-and-click interfaces. They are also some of the best tools for making dashboards:

- Tableau
- Power BI
- Qlik
- AWS QuickSight

These are all useful tools, and as with spreadsheet tools, if you know one of them well, you can figure out any of them.

Business services

These are different than the other tools. Instead of being a piece of software, these are companies that offer data services:

- Apex Systems
- Dataroma

It's debatable whether these should be on the list, but the fact of the matter is that they are. That means these names may pop up in the exam, and you need to be able to recognize them.

All-purpose tools

This category is for software that is designed to be a neat package that will solve all of your data problems (ha!):

- RapidMiner
- IBM Cognos
- IBM SPSS Modeler
- IBM SPSS
- MicroStrategy
- BusinessObjects
- Domo
- Minitab

These are mostly enterprise-wide software solutions that a company buys a license for and certain people within the company get to use.

Which tools you should learn to use

This is not part of the exam, it is just my opinion, so feel free to skip it if you wish to. Whenever I teach this section, people always ask which tools they should learn to use. This is another one of those topics that people will debate until they turn blue in the face, sure that their tool of choice is the only one worth knowing. The truth is not quite that simple. Almost all data analysts will need to know how to use multiple tools. There are a few positions that will only require a single tool, and they want a specialist who has mastered the heck out of that one tool.

The correct answer is that you should learn about the tools that are in use in the specific field you want to enter. If you look up lots of job postings for your dream position, chances are a lot of them will use the same few sets of tools. For example, military contracts use a lot of SAS, some hospitals use R, academia uses a lot of SPSS, a lot of tech companies use Python, and some businesses only want Power BI. Different fields will have drastically different tool sets, and you need to develop a tool set for the job you want. It doesn't matter whether you are a master of R if the job you want requires Tableau. Figure out which types of positions you want, and then figure out which tools they are using in the field right now.

Okay, that was the technically correct answer that no one wants to hear. What should you learn if you don't know which field you want, or you just want to be a data analyst and don't care about the field or the company? Personally, I recommend you learn one query language, one programming language, one spreadsheet software, and one visualization software. For example, you might want to learn MySQL, Python, Google Sheets, and Power BI. This will make you a well-rounded data analyst; even if you don't know the specific tool the job wants, chances are you know something like it and can learn it easily enough. There are lots of resources available for learning how to use them, and these are all either free or have a free version that you can use to learn.

You will note that I did not include any of the all-purpose tools here. These tools are expensive, and most companies will train you in them if they want you to know them. In my opinion, unless the job of your dreams requires you to know one right out of the gate, they aren't worth the bother.

Summary

This chapter has been pretty straightforward. First, we covered static and dynamic reports, what they are, and the pros and cons of each. Then, we talked about ad hoc reports and research reports, one being a short, quick report and the other being a long, in-depth report. Next, we discussed self-service reports, what they are, and why they are popular. Then, we briefly covered recurring reports and some common types. Finally, we wrapped up this chapter with a brief summary of a list of data analytics tools you will need to know the names of for the exam. This has been a high-level overview of the different types of reports.

In the next chapter, we will cover what goes into these reports!

Practice questions

Let's try to practice the material in this chapter with a few example questions.

Questions

1. A bank, *Second Rational Bank*, requires all of its tellers to have access to metrics such as the interest rates and fees on savings accounts, which change regularly. Which type of report would be most appropriate for this need?

   A. Point-in-time report

   B. Real-time report

   C. Research report

   D. Ad hoc report

2. *Second Rational Bank* is considering buying a package of loan debts from a sister company. It wants to know what the risks are, how much it can offer the other bank while still making a profit, and what it will need to do after to collect on the loans. Which type of report would be most appropriate?

   A. Self-service report

   B. Ad hoc report

   C. Recurring report

   D. Research report

3. A manager at *Second Rational Bank* wants to track the metrics of their team and make sure they are meeting all rules and regulations of the company. Which type of report is most appropriate?

   A. Recurring report

   B. Research report

   C. Ad hoc report

   D. Self-service

4. Which of the following is not a data analytics tool?

   A. SAS

   B. SPSS

   C. STS

   D. Stata

5.   Which of the following data analytics tools is specifically designed for visualizations and reports?

   A.   Power BI

   B.   Microsoft Excel

   C.   SAS

   D.   Domo

Answers

Now, we will briefly go over the answers to the questions. If you got one wrong, make sure to review the topic in this chapter before continuing:

1.   The answer is: *Real-time report*

   The bank tellers need quick responses with the most up-to-date information and little to no complexity. This is an ideal time to use a real-time, or dynamic, report.

2.   The answer is: *Research report*

   This is a big question with a lot of parts to it and can mean a gain or loss of millions of dollars to the company. The ideal solution is to take your time and write up a full research report.

3.   The answer is: *Recurring report*

   Specifically, this would be a compliance report, which is a type of recurring report, because the goal was to check metrics against requirements and regulations.

4.   The answer is: *STS*

   Yes. you can create visualizations and reports with any of these tools, but Power BI is specifically designed as a visualization and reporting software. Make sure you learn not only all of the tools on the list but the group that gives them a general purpose.

5.   The answer is: *Power BI*

   Yes, you can create visualizations and reports with any of these tools, but Power BI is specifically designed as a visualization and reporting software. Make sure you learn not only all of the tools on the list but the group that gives them a general purpose.

12
Reporting Process

Being able to effectively communicate your findings is one of the most important skills a data analyst can have. More often than not, the results of all of your hard work will be communicated through a report. However, you can't just slap some numbers in a PowerPoint presentation and call it a day.

In this chapter, we will discuss the process of creating a report. Then, we will cover the things to consider while creating a report. Next, we will talk about the different elements of a report. Then, we will discuss what to think about while delivering a report. Finally, we will cover how to make your reports look professional.

In this chapter, we're going to cover the following topics:

- Understanding the report development process
- Knowing what to consider when making a report
- Understanding report elements
- Understanding report delivery
- Designing reports

Understanding the report development process

Let's talk about the process of actually creating a report. The process may be slightly different depending on the procedures of where you end up working, but this is the general process identified by the experts at CompTIA. For the exam, you will need to know the steps involved in the report creation process, have a general idea of what they mean, and know what order they should be done in. I will point out that for the exam, this process is specifically related to dashboard development, but the concepts have been generalized here to cover all reports Whether we are talking about a dashboard or any other kind of report, except for an ad hoc report, the process will be roughly the same. Ad hoc reports generally have a less formal process because they require a quick turnaround time. The four-step process is as follows:

1. Create a plan.
2. Get the plan approved.

3. Create the report.

4. Deliver the report.

If we want to see this in a visualization, it would look something like *Figure 12.1*:

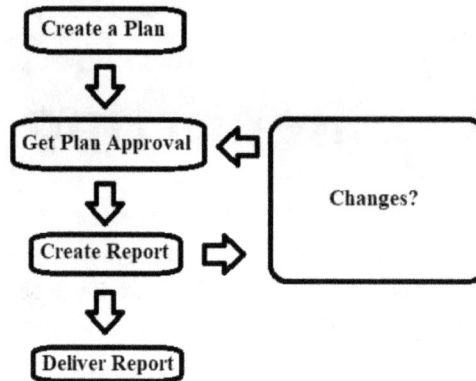

Figure 12.1 – Report creation flowchart

Figure 12.1 does include an option to make changes while you are creating the report. If in the process of making a report, you realize that changes are required, it is important to go back to the approval process. Even if the result is better than what was originally planned, a lot of bosses will be upset if you don't give them the report you promised them. This is fairly straightforward, but let's go through this step by step.

Creating a plan

This step actually can be called several different things and may go by different terms on the test. These terms include mockup/wireframe, layout/presentation, flow/nav, or data story planning. Some of these are exclusive to a dashboard. Mockups and wireframes specifically refer to creating a blueprint for how you want the dashboard to look. However, layout and presentation effectively refer to the same thing in a more general sense. These are all about how you want to physically present your data in a report. Flow, nav, and data story planning are more about the order in which you present your data and how that impacts the message you are getting across to your audience. To be more specific, flow is the order in which you present your findings, nav is how you allow your audience to navigate through the dashboard, and data story planning is the final picture of what this all means.

Let's take a moment to talk about data story planning. This is one of the most important skills a data analyst can have, and it is highly underrated. It doesn't matter if you have come up with an equation that can magically end world hunger if you can't explain it to the people who can make the change. The idea is simple: a data story is what you can communicate with your data. However, this step is often skipped, and many analysts have no idea what they are actually communicating or whether it is in line with what they have actually found. Data stories are not covered by the exam, besides that in the most general sense, they are part of the planning process, so we won't go into further detail about them here. However, if you are not familiar with them, I suggest you look into them when you have some free time. There are several useful sources, but the book *DataStory: Explain Data and Inspire Action Through Story* by **Nancy Duarte** is a popular one.

Back on topic, no matter the term used, this is the step where you get your ducks in a row, figure out what you need to answer the question, and how you want to present the results.

Getting the plan approved

This step is more relevant to dashboards and research reports but can be applied to anything. The idea is that you take the plan you just created to your boss or the person who asked the initial question and make sure that what you are planning is what they actually want or need. This is an important step because the processing of creating these reports can take a fair amount of time, time for which the company is paying you, not to mention the extra expenses if you have to run a study or buy data from a third party. Imagine if you spent all of that time and money to generate a report and it wasn't what they needed at all.

This step is to double-check that you and the person for whom you are generating the report are on the same page. Not every company will make you do this step, but it is best practice. The downside to this step is that once it is approved, you need to deliver specifically what was approved. You will need to go back to the person to get any major changes approved, or else you are not delivering what you were paid to create. To be clear, you are not getting a specific result approved, just checking what specific questions are being answered and how those results are presented.

Creating the report

This is pretty self-explanatory. You made a plan, you got the approval, and now you just have to make the actual report. Gather your data, clean it, wrangle it, analyze it, and fit the results into a nice, easy-to-understand report. More specific steps depend on the kind of report and what you are reporting, but the general process is the same, whether or not it is a dashboard.

Delivering the report

Finally, you have to give the report to the person who asked for it. For a dashboard, this is when you deploy it to production, but for other reports, this is often where you present your findings at a meeting, or, in a more casual context, explain the results over a nice cup of tea. Sometimes, it is just a file or even a piece of paper that you pass on to someone else. How the report is delivered will depend on the report, who requested it, and how they want to receive it, but this is an important step. Answering questions is meaningless if you don't tell someone what you found. There are a lot of things you have to have to consider when it comes time to deliver your report, but we will cover these later in the *Understanding report delivery* section.

That's it for the process: create a plan, get the plan approved, create a report, and deliver the report. This is, of course, easier said than done, but for the exam, you only need to remember these steps. Next, we will talk about the things we need to consider when making a report.

Knowing what to consider when making a report

In this section, we will cover things that you need to take into account as you make these reports. This seems kind of vague, but these are concepts that you are expected to know for the exam and about which you may be asked in the context of a report-generating scenario. In other words, they will give you an example situation in which you are generating a report and ask how you would handle specific aspects depending on the situation. This is a part of the exam that confuses a lot of people. Not only do you need to understand these concepts but you also need to be able to apply them appropriately in a given scenario. These questions are designed to test experience, but you can figure them out if you take your time and think through them carefully. Don't panic. Even if you don't have experience, you *can* do this. Let's jump right in.

Business requirements

The first list of things to consider is business requirements. CompTIA specifically says, to *"translate business requirements to form a report."* What they are saying is that you need to take some given information and explain what it practically means. Let's start with the list of business requirements you are asked to translate:

- Data content
- Filtering
- Views
- Data range
- Frequency
- Audience

These concepts are not so bad – you are already familiar with most of them. The trick is knowing how to apply them. Let's start at the top.

Data content

The data content is exactly what it sounds like: the variables, tables, or visualizations you include in your report. This can be trickier than it sounds and interacts with things such as its audience, which we will discuss in a little bit. The idea is that you need to include the information required to answer the question, but nothing extra. The more things you add that you don't need, the more convoluted and muddy the report becomes and the less likely your audience will be to understand it.

There are two things you need to consider when thinking about data content. The first is what variables you require to answer the question, and the second is making sure you don't have anything you don't need. This can take many forms on the exam, including being given a dataset and a question and asking which variables should be included in the report. They may also simply ask a question, then each answer is a different combination of variables. Missing a variable that you do need or including a variable you don't need will both be considered wrong answers in the exam.

Let's go over a brief example. There is a team in a robotics company whose *only* goal is to make their robot faster. They want a recurring report every week to see whether they are making progress and to make alterations and plan for the next week. Which variables from the following table in *Figure 12.2* should be included in the report?

(Note: the equation for speed is speed = distance/time.)

Date	Time(s)	Vol	Ins	Res	Operator	Distance(m)	RobotWeight(lb)
7/5/2019	33	1.9470	471	40	Oscar Mayor	22	9
7/12/2019	35	1.3958	581	86	Oscar Mayor	25	11
7/19/2019	44	1.3905	304	4	Oscar Mayor	23	10
7/26/2019	30	1.5938	482	15	Oscar Mayor	24	8
8/2/2019	43	1.1938	561	51	Oscar Mayor	19	11
8/9/2019	33	1.8404	309	27	Oscar Mayor	21	9
8/16/2019	35	1.4893	701	83	Oscar Mayor	16	9
8/23/2019	40	1.2843	824	79	Oscar Mayor	15	10
8/30/2019	35	1.9837	754	73	Oscar Mayor	22	11
9/6/2019	43	1.7489	601	48	Oscar Mayor	24	9

Figure 12.2 – Example dataset

Here, we can see a lot of data is generated by these runs, more than we need. Still, there is no variable for speed in the raw data. Since all the company wants to know is whether their robot is faster than last week, you only need to include three variables in your report: distance, time, and date. You will use distance and time to calculate speed in the report, so you need both of them. The tricky part that catches most people is that you need the date variable so that you know which data is from this week and which data is from last week. There is no way to compare results from one week to the next without something to distinguish them. In this case, this is the date variable. Alternatively, it could have been a version number, or trial number – something to break the data down into groups so that you can compare them.

Working as an actual analyst, you will probably be asked to track many more things, and there is no penalty for including an extra variable here or there, but the test has to be strict. Consider not only which variables you will be comparing but also the variables that you will logically need in order to compare them.

Filtering

We have discussed filtering in previous chapters. There are a few different ways to interpret this. The first is simply that you should use filters so that you only have the data content you require. The other is specific to dashboards or self-service reports, where you can add the ability to filter. Making a dashboard interactive and giving the audience the ability to filter specifically what they want to see means that they can get more value out of the dashboard, but it does slow the dashboard down, so this should be used sparingly.

If we are looking at a classic static report, filtering mainly happens at two different stages:

- Query filtering
- Report filtering

Query filtering is when you are first pulling the data to your local environment to use, and report filtering happens when you are figuring out specifically what you are going to use in the report. You may also filter or subset the data during the data-wrangling process, but that depends on the specific things you will be doing. You are much more likely to have a filter at the beginning of the data process and the end.

If the exam question is talking about a dashboard, then they are probably talking about a filter as an interactive element. If the question is about something such as an ad hoc report, then it is talking about a filter to narrow down your data content for the report, or what variables are represented in your visualizations or tables.

Views

Views are functionalities that exist in a couple of different data tools, but mainly SQL or Power BI. Effectively, this saves the result of a query as a separate data object. A view can be visualized or queried just like any real table. They are, in fact, quite similar to a temporary table, but they don't delete themselves and they are **READ-ONLY**, which is important, but we will come back to that in a moment. There are two reasons to generate a view:

- Cleaner code
- Data integrity or security

First, we will talk about cleaner code. Just like a temporary table, if you have a query that is over 100 lines of code long, it is just easier to save it as its own data object so that you don't have to use a very long, very slow query every time you want to check something. This is faster, more efficient, and easier for someone reading your code to understand.

The main reason views are used is to protect the data. Sometimes, there will be someone who has just enough knowledge of data to be dangerous. They took a class in high school and now they want to look at the data themselves, but if they accidentally delete half of the database that the company has spent millions of dollars on, it is you who is going to be in trouble. Deleting data is probably the best thing they could do and can usually be recovered. From experience, it is strangely easy to accidentally delete millions of records. Don't ask. More importantly, they could alter values without anyone noticing – then, every analysis everyone in the company ran using that data would just be wrong. To keep this from happening, you can create a view and give them access to it. This way, they can look around the data to their heart's content without actually posing a risk to anything because views are READ-ONLY and cannot be used to change the original values.

A quick example might be Steve from HR, who is not a data professional but has requested access to the data because there is something he wants to look up for himself. This is the perfect time to use a view.

Data range

The data range is similar to data content and even relates to the audience, but instead of what you are including, it pertains to how much you are including. This can be applied to any number of things, such as date ranges, geographical ranges, department ranges, or any other range in your dataset. Just like with data content, if your data range includes too much or too little, then you will get the question wrong.

In general, the more specific a question is, the smaller the data range will be, and the vaguer the question is, the larger the data range will be. If it helps, you can think of it like a seesaw, as in *Figure 12.3*:

Figure 12.3 – Low specificity, high data range

Here, we can see that when specificity is low, the data range is high. Now, let's see the opposite in *Figure 12.4*:

Figure 12.4 – High specificity, low data range

Here, the specificity is high, so the data range is low. There is also some middle ground, as in *Figure 12.5*:

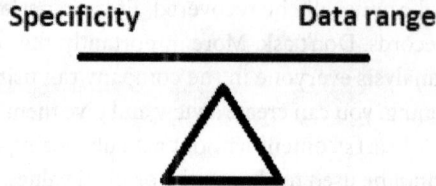

Figure 12.5 – Moderate specificity and data range

Here, in *Figure 12.5*, both the specificity and data range are moderate. To be clear, there are exceptions, but this is the general rule. None of these are better or worse; they are just used in different situations.

This can be a little confusing, so let's look at some examples. A hospital has implemented a new software for receptionists to use to schedule appointments. After tracking the time it takes for each receptionist to schedule every appointment, they want to know whose average scheduling time is decreasing, whose is staying the same, and whose is increasing so that they can offer additional training for the new software to those who are struggling. What is the most appropriate data range to use for time: day, week, month, or year?

This is highly specific – we are looking at the time it takes a very specific group of people to do a specific task using a specific tool over a very short period of time, so this will have a very low data range. The confusing part is that we are looking at data over the course of the week, but we are actually going to use days to do it. Here, the correct answer is days because we have to break it down to days so that we actually have more than one value to compare.

After a couple of years, the person asks whether the new software was faster than the old software. This is very low specificity, so you would have a much higher data range.

In general, for these questions, take your time and think through what the person asking the question needs and what you need to figure out.

Frequency

Frequency is how often a recurring report is given and is linked directly to the purpose of the report. Is the report given every year? Quarter? Month? Week? Day? Hour?! In an ideal world, the unit of time between recurring reports should be one unit of time smaller than the goal you are trying to reach. If you want to meet a yearly goal, check quarterly; if you want to meet a quarterly goal, check monthly, and so on. This rarely happens in the real world, because few companies are actually this organized. They check their quarterly goals quarterly and their monthly goals monthly. This means that they either pass or fail and have no control over it. They just try not to fail the next one too.

Luckily, in the exam, you don't really have to worry about this. Often, you will be told in the question that they want the report weekly and then they will ask what the frequency is. Alternatively, they can say something such as "a project manager wants a report to plan for their next 1-week sprint," and ask what the frequency is. On occasion, they might say that they want a couple of reports a month, and ask for the best frequency. In this scenario, just pick the frequency that makes logical sense based on the question. Try not to overthink it.

Audience

The audience is an important factor simply because it impacts every other aspect of the report. You can think of the audience as the people who will receive the report, which can mean a distribution list for a dashboard or an email, or those in attendance at a meeting. Whoever is intended to receive the report you are writing is the audience.

The guideline for the audience is never to give a report to someone who doesn't need it. There are several reasons for this: you are wasting everyone's time, causing unnecessary confusion, can leak confidential information, and there are legal issues if you are using protected data.

Let's take a moment to consider an example. A manager in the sales department requests information on how a specific variable – let's say customer demographic information – impacts the sales of a given product. This is probably a report that would go to all of the managers in the sales department and, perhaps, the head of the sales department who is directly above these managers. This audience is the level of people who are directly influenced by this information. They are the ones making decisions in the sales department based on what you find. Giving this data to people below this level, above this level, or at this level from a different department would be inappropriate. The exception to this is if someone at a higher level specifically requests the information, or that the information can be distributed.

Always consider your audience first. Think of what they need, how much of it they need, and who, exactly, needs to know. Now, this is looking at the audience in the broad sense, and there may be questions about that, but the exam goes on to break the audience down into specific consumer types, so let's make things a little more granular:

- C-level executives
- Management
- Stakeholders
- General public
- Technical experts

These are general audience types identified by CompTIA called Consumer Levels, and they will each be treated differently. Let's start at the top. C-level executives are those with job titles that start with Chief. These include **Chief Executive Officer (CEO)**, **Chief Financial Officer (CFO)**, **Chief Information Officer (CIO)**, **Chief Human Resources Officer (CHRO)**, **Chief Marketing Officer (CMO)**, **Chief Operating Officer (COO)**, **Chief Technical Officer (CTO)**, **Chief Snacking Officer (CSO)**, or anything else you can think of that starts with "chief" and ends with "officer." The point is that these are all high-level executives in the company. C-level executives have a high level of access and receive information at a high level. Let's be clear about what that means. A high level of access means that they are allowed to hear about confidential information pertaining to the company; they run it, so they can hear what they want. Receiving information at a high-level means that they only receive broad summaries, usually with a large data range. They don't care about days or weeks – more about how the company is doing from year to year.

The next consumer level is management, which includes everything from team leaders to middle managers. Management has mid-level access and receives mid-level information. Here, they do not have access to everything, but they have access to a fair amount – usually, everything pertaining to their team or department, but not other departments. Mid-level information means that their information is still summarized, but not as much and they may have some granular information. The data range here is on a smaller scale than the C-level executives and data is usually shared on a monthly or quarterly basis.

Stakeholders are people who have a vested interest in the company. There are many different kinds of stakeholders, including employees, customers, suppliers, communities, governments, or trade associations, but the most common type of stakeholder is an investor. Maybe they own stock in the company, or maybe they just gave it money, but they often want occasional reports to know that their investment wasn't a waste. Stakeholders, in general, have mid-level access but receive information on a high level. They will generally not know company secrets, but they will have access to enough information to see how the company is doing. The information that they receive is highly summarized and often includes a broad data range. They don't care about nitty-gritty details; they just want to know whether they will make money or lose it.

Technical experts are, in my completely unbiased opinion, the best. Note that this is where data analysts fall. Technical experts are a diverse bunch, including subject matter experts, specialists, and contractors. These are the people who focus on doing one thing very well. They receive low to mid-level access, and they receive data at a very low level. In terms of access, it depends on the position; some positions need very little information. An engineer who assembles circuit boards doesn't need a lot of data besides the readings from the board they are actively working on. A networking engineer who manages the connectivity of an entire branch of the company requires considerably more access. However, technical experts almost never have a high level of access, unless they reach a position such as CTO, which is both technical and C-level. The information technical experts receive is often very detailed and granular. This is the level most likely to just get raw data, and they are also the level that is supposed to know what to do with it. The data range for this group is small. They may require information for weeks, days, hours, or even smaller units of time.

The general public includes anyone in the world not affiliated with the company. This is a miscellaneous "everyone else" category. Here, they will have low to no access, and if they do receive any information, it is at an extremely high level. Companies will often announce things like "we are growing" or "we made money this year." These reports are vague and largely meaningless. Information that is released to the general public is highly scrutinized and monitored. When in doubt, tell the general public nothing.

Okay – we have covered the audience in a broad sense, as well as specific consumer levels. Let's move on to considerations that are specific to dashboards.

Dashboard-specific requirements

There are some considerations when making a dashboard that don't apply to other report types. All of these are focused on how the data relates to the dashboard itself and can be broken down into two main categories:

- Data sources
- Data attributes

Some of this information will be similar to other reports but called something else, and some of it will be new. Let's jump right in!

Data sources

When we are talking about data sources, we are talking about two things. First, what is the actual database, or what are the databases, from which we are getting this data? Second, how is that data connected to the dashboard? Let's assume you know what database you are using and focus on the second one. There are three types of connections:

- Live
- Last extracted
- Embedded

The first two should sound familiar! A live connection means that it is a dynamic or real-time report. The last extract is the same as a static report, with a date to say specifically when the data was last refreshed. The first two connections are considered reusable, which means that the data source can be used multiple times for multiple different reports, but the third type is different. Embedded means that the report is actually embedded into the data. That data source can only be used with that report and they are published together.

Data attributes

Data attributes are just renaming things. Why do dashboards feel the need to use different words to define things that already exist, making an already muddy lexicon even more confusing? It is a great mystery of the universe. In any case, here are a few words you need to know for the exam:

- Dimensions
- Measures
- Field definitions

Dimensions are qualitative, or categorical, variables. Whatever you want to call them, these are the variables that break things down into groups. Measures are quantitative, or numerical, variables. They hold counts, measurements, and calculations. All variables will automatically be sorted into these two categories, but you can change them manually if, for example, you have an ordinal variable that can look like a number but is actually used to create groups.

Field definitions give specific information about a variable. Every variable has a field definition and it usually includes information such as the data role, data type, and domain, how the data is aggregated, and the formula if it is a calculated variable. You do not need to memorize what specific things a field definition will record because it changes from software to software. All you need to know is that field definitions describe a variable.

Okay – let's bring this together now. When you are making a dashboard, you need to consider where your data is coming from, how it will be connected, what data to include, how much data to include, and the type of consumer your audience will be. Easy, right? Except that's not all. Oops! CompTIA also includes a list of things to consider specifically when you are delivering a dashboard, and we will talk about that in just a minute. Before we talk about delivery considerations, we need to think about what actually goes into these reports.

Understanding report elements

There are lots of things that can go into a report besides the actual data. These may serve different purposes depending on how and when they are used, and you might not always need all of them. In the exam, they might give you a scenario or a picture of a report and ask you which elements are missing or should not be there, or they might give you a scenario and ask which elements need to be updated. Let's go over them one at a time and discuss why they are important. First, here's a list of elements:

- Cover page
- Version number
- Reference data sources
- Reference dates
- **Frequently Asked Questions (FAQs)**
- Appendix

The first thing you see on a report will be a cover page because the cover page is, literally, the first page of the report. If your report is a dashboard, then the cover page will provide instructions on how to use the dashboard. It will give a brief rundown of the different elements, any interactive parts, and generally, how to navigate it. If your report is not a dashboard, then the cover page may contain a summary of observations and insights. In the military, this is called the **Bottom Line Up Front (BLUF)**, and as soon as someone picks up the report, they know why it is important and why they should bother reading further. Now, the cover page is slightly different from the other report elements because it will actually contain other elements. You will often have the version number, reference dates, and sometimes the reference data sources right upfront on the cover page. You can see an example of a cover page in *Figure 12.6*:

```
┌─────────────────────────────────────────┐
│                Cover Page                 │
│                                           │
│              Report Title                 │
│            Version Number                 │
│                                           │
│        ┌───────────────────────────┐      │
│        │                           │      │
│        │                           │      │
│        │     Summary (Static)      │      │
│        │   Instructions (Dynamic)  │      │
│        │                           │      │
│        │                           │      │
│        └───────────────────────────┘      │
│                                           │
│                                           │
│  Data Source                              │
│  Data Refresh Date                        │
│  Report Run Date                          │
│                                           │
└─────────────────────────────────────────┘
```

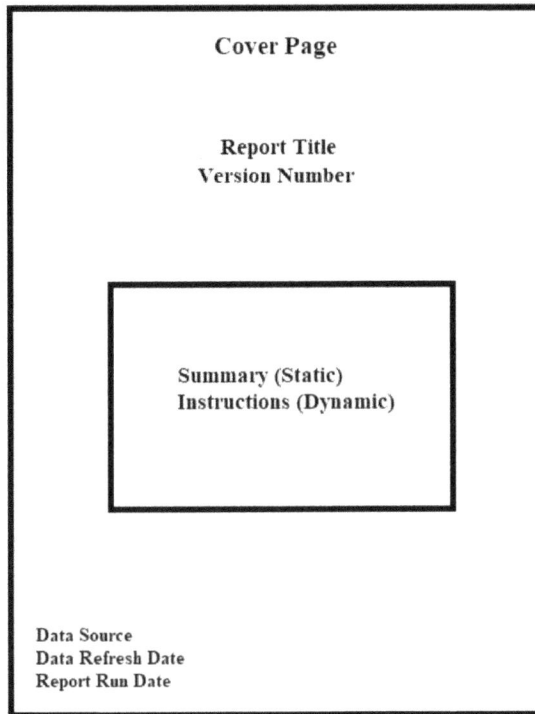

Figure 12.6 – Cover page example

Again, there is more than one way to format your cover page, but this is a common layout.

Not all reports have a version number because not all reports need them. Only reports that get updated or changed will have more than one version and require a version number. Imagine an ongoing project and the report for it gets updated with every new discovery or change in the project. It doesn't make sense to change the name and write a new report from scratch every time. Instead, you change only what you need to and give the report a new version number. That way, everyone can make sure they are talking about the same version of the report. Every time a report is updated, it should get a new version number.

Reference data sources are what you would guess – they simply denote what data sources were used in the generation of the report. They should be on pretty much every report because you should always be using data from somewhere. That said, this rarely changes, unless you get new data from a new source.

Reference dates usually refer to two specific dates:

- Report run date
- Data refresh date

The report run date is the date that you actually generated the report in the case of static reports. This date doesn't get displayed on a dashboard because, for a dashboard, the report run date is the last time the report was used, so if you are looking at it, it will always be the current date. The data refresh date is the last time the data was pulled, so it only applies to static reports. In other words, a dynamic dashboard would have neither of these dates, a static dashboard would have a data refresh date and no report run date, and any report that isn't a dashboard will probably have both.

FAQs is a section at the end of the report that answers questions you have received about the report multiple times. If you have already received questions multiple times, that means that it is either a self-service report or a recurring report. You do not generally find these in ad hoc or research reports. This section is updated whenever there is a question that you are sick of answering. If there is an element of the report that continues to confuse people, you can either add it to the instructions on the cover page (if it is a dashboard) or add it to the FAQ section at the end.

The appendix is where you shove any additional information that relates to the report, but is not necessary to understand it. If you must include your code, and I don't recommend it, it would be in the appendix. You can also include things such as definitions or regulations that can add value if the reader is interested, but that are not required.

Let's consider an example. You work for an apple orchard and have created a static research report that explains orchard productivity. The company is doing so well that it buys a smaller competitor, and you now have to create a new report that includes data from this new source. Which elements of the report will change?

The cover page, the version number, the reference data sources, the report run date, and the data refresh date. You will have new observations and insights, and the other elements change, so the cover page will be updated. This is an old report that is being changed, so it will get a new version number. You are adding an additional data source, so the reference data sources will change. Finally, you have to pull data again because you are also pulling it from a new source, so the data refresh date is changed, and you are rerunning the analyses to include the new data, so the report run date changes. This is new information, so there should be no changes to the FAQs or the appendix unless the question specifically tells you there is something to add.

Now that you have a pretty good idea of what types of things go into a report and how they influence one another, we can move on to what you have to consider when you are delivering a report.

Understanding report delivery

This section is listed as report delivery, but the majority of these considerations apply directly to dashboards. CompTIA provides of list of delivery considerations:

- Subscription
- Scheduled delivery
- Interactive saved searches

- Filtering
- Static
- Web interface
- Dashboard optimization
- Access permissions

This list is kind of confusing, so we are going to roll these up into some general groups:

- When it is delivered:
  - Subscription
  - Scheduled delivery
- How it is delivered:
  - Web interface
  - Access permissions
  - Static
- How optimized it is:
  - Interactive saved searches
  - Filtering
  - Dashboard optimization

First, when it is delivered – a subscription is a way to schedule a dashboard to automatically be refreshed and sent to selected recipients through a medium of your choice, but usually, email or Slack. For example, every Monday at 7:00 A.M., it will refresh and send out an email to everyone on the list to connect to it. A scheduled delivery is very similar but only happens once. If you want to do a scheduled delivery every Monday at 7:00 A.M., then you need to set up each one individually. Besides that, these are effectively the same thing.

Next, how it is delivered – in this case, using a web interface just means that you have your dashboard hosted on the web, usually by a third-party company or software service. Alternatively, you can host it locally if your company has a local network. You can think of this as deciding what you are hosting the dashboard on. Access permissions are just a security feature so that only the intended audience can see the dashboard. You can imagine that if you are posting confidential information on a dashboard that you are hosting on the web, you will want to make sure that only your intended audience gets to see it. Static, in this case, is referring to how your dashboard is connected to your data, statically or dynamically, which we have covered multiple times already.

Finally, how to optimize it – saved searches and filtering are both ways to make your dashboard more interactive. The audience can click on more things to change the visualizations. How interactive you make it will often correspond to the level of information received by your consumer level. A low level means you are working with technical experts and you can afford a high level of interactivity because they know what they are doing. Mid-level means you are working with managers, and you can add a moderate amount of interactivity and complexity. High-level means you are working with C-level executives or stakeholders, and you will have little to no interactivity. The flip side of this coin is dashboard optimization, which focuses on how fast and efficient your dashboard is. There are a lot of things that play into dashboard optimization, but one of the main things is how complicated it is. The more interactive elements you add, the slower it will be, so it is a balancing act, and you should really only add interactive elements if the audience will directly benefit from them.

That is everything you need to know about report delivery for the exam. Next, we will cover designing a report.

Designing reports

This section is all about the design elements of a report. This is something that is often ignored, and several people have wondered why it is in the exam, but I promise it is important. A lot of these things are to make your reports cleaner and easier to understand, but they also make your reports look more professional. Even if you are a great data analyst, if your reports look like a two-year-old hopped up on sugar went hog-wild with an entire box of crayons, no one will take you or your report seriously. The sad truth is that you will often be judged by the polish of your work more than its content. Let's figure out how to make your reports shine!

Branding

Before we talk about anything else, we will talk about corporate reporting, specifically branding. Some companies, especially larger ones, will have a brand, or a specific image they want to project. That means that they may have color codes, logos, watermarks, fonts, layouts, or even full templates prepared. Branding guidelines are more important than anything else, and they trump anything else you will learn about in this chapter. If your company colors are neon pink and burnt orange, then you should use those colors in your report, in your header, and even as the colors for your charts and tables. It doesn't matter if they are ugly as sin – if that is company policy, then you make those ugly charts.

Yes, this is painful, but always use brand guidelines first. Check to see whether there is a logo or company color scheme and apply them to your reports. A company may also have different guidelines for different consumer types. For example, a report released to the general public will probably look different than a report made for a technical expert.

Fonts, layouts, and chart elements

Okay – fonts, layouts, and chart elements are small things that will make a big difference in how professional your reports appear. Let's start at the top.

Fonts

There are several things you need to consider when you are thinking of fonts. No, we are not going to talk about advanced typography, but it is an interesting topic if you have nothing else to do. Instead, we are going to lay down some guidelines:

- Clarity
- Professionalism
- Number of fonts

The first rule of fonts is that your audience needs to be able to read them. There are lots of things that impact clarity, including font size, line spacing, and color contrast. Font size is pretty obvious; if the font is too small, no one can read it. Try not to go below font size 11. Anything below that can be problematic, as seen in *Figure 12.7*:

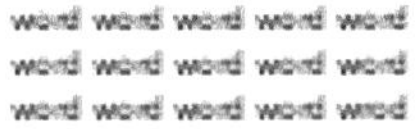

word word word word word
word word word word word
word word word word word

Figure 12.7 – Tiny font

Next is line spacing. There are no hard rules about this, but if your lines are too close together, it makes them difficult to read. You can see this for yourself in *Figure 12.8*:

word word word word word
word word word word word
word word word word word

Figure 12.8 – Line spacing

Finally comes color contrast. This is the color of the font against the color of the background. There should be some contrast, but not too much. If you have too much contrast, it physically hurts the eyes and tires them out – no one will thank you. You can see this in *Figure 12.9*. Note that there are better examples of this, but not in a book that is printed in black and white. Try putting a neon green font on a bright red background if the following example doesn't work for you:

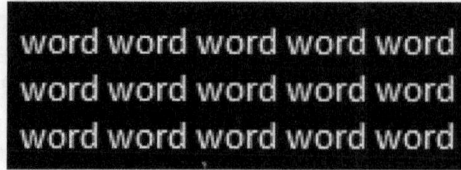

Figure 12.9 – High contrast font and background

On the other side of the spectrum, if the contrast is too low, then it makes the text difficult to read, as seen in *Figure 12.10*:

Figure 12.10 – Low contrast font and background

The ideal contrast is somewhere between these two extremes. If you are ever in doubt, black font on a white background will work just fine. That said, try to avoid a white font on a black background.

Okay – so if you have a reasonably sized font, with reasonable spacing, and reasonable colors, you have clarity. Next, let's talk about professionalism.

Certain fonts are frowned upon. Let's take a look at a few now:

Figure 12.11 – Unprofessional fonts

All of the fonts like those in *Figure 12.11* are to be avoided unless they are part of the company brand. There are a few fonts that are more popular and generally considered professional-looking fonts. Let's see a few of them now:

Calibri

Cambria

Times New Roman

Figure 12.12 – Common professional fonts

In *Figure 12.12*, we can see some fonts that are generally considered professional: Calibri, Cambria, and Times New Roman. Of course, other fonts are acceptable, but you will notice that all of these fonts are clean and easy to read.

That brings us to the number of fonts. As a rule, you should never more than two or three different fonts. Lots of different fonts make the report look cluttered and confusing. This may seem nit-picky, but it will be on the exam. You might be shown a picture of a report with five different fonts and be asked to explain what is wrong with it. Keep it clean, keep it simple.

Layouts

If you are making a text document, a slideshow, or a dashboard, how you lay out your report does matter. We have all seen slides with way too much text, and they are difficult to read, especially if you are trying to listen to the person giving the presentation. This leads us to the most important rule of layouts. Don't overfill. White space is your friend, believe it or not. There is no need to cram as much as you possibly can into the smallest space possible. Generally speaking, you should only have one visualization per slide or page of your dashboard. This is part of communicating a data story effectively. When you try to cram too much into a small space, it makes all of it harder to absorb. To that same end, you should only have one data story per visualization. It might be tempting to have one omnichart that answers five different questions, but they are difficult to read and confusing.

This isn't in the exam, but as a general practice, never write your entire script on the slide. A slide should only have key points and it is up to you to remember what to say about each of them. This will help you avoid having huge blocks of text on a tiny slide.

If you want to go above and beyond, you should learn about the rule of thirds. Again, this is not in the exam, but if you break the slide into a 3 x 3 grid, the places where the lines intersect are natural focal points. You can see this in *Figure 12.13*:

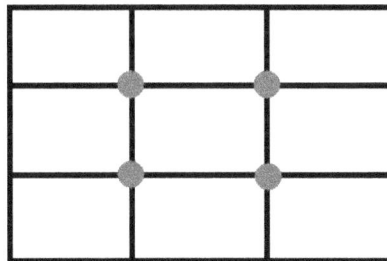

Figure 12.13 – Rule of thirds

This is interesting and can take your reports to the next level visually, but it is by no means required to create an effective report. The same rules apply here. Keep it clean, keep it simple.

Key chart elements

There are three keep chart elements:

- Titles
- Labels
- Legends

The title is what you call the chart, labels identify things such as axes, and the legend is a table that identifies other chart elements. You can see these in *Figure 12.14*:

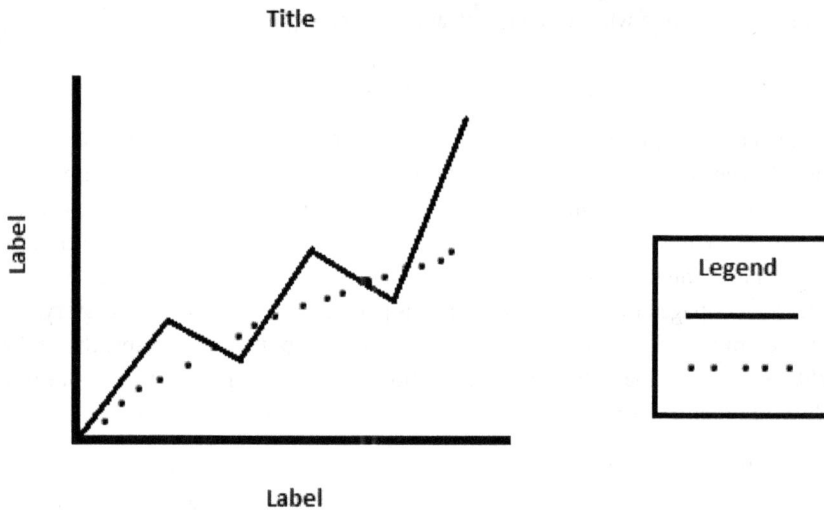

Figure 12.14 – Title, label, and legend

For the exam, you will need to be able to distinguish between these. You will also need to know what makes a good title, label, or legend. These should all be accurate and concise. In *Figure 12.15*, we can see a chart, and then we have a list of possible titles for this chart:

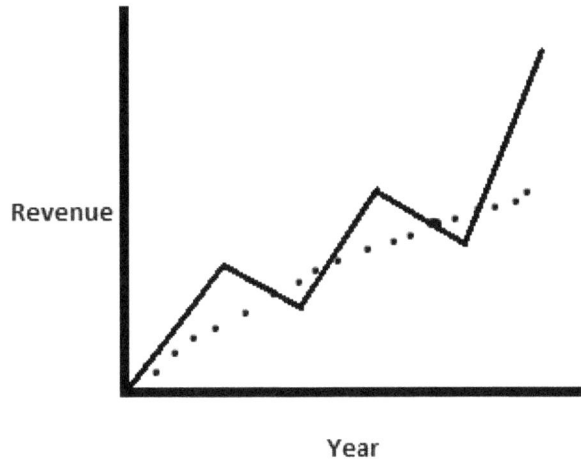

Figure 12.15 – Example chart

Possible titles:

- **Revenue**

- **Revenue from 2010 to 2020**

- **Number of Sales for Each Year Multiplied by the Average Price of Service per Year for the Years 2010, 2011, 2012, 2013, 2014, 2015, 2016, 2017, 2018, 2019, and 2020 as Aggregated from Quarterly Reports**

- **Cost of Living per Region**

Which title do you think is most appropriate? The correct answer is **Revenue from 2010 to 2020**. This states exactly what the chart includes clearly and concisely. If you were to just call the table **Revenue**, it would become too vague and you are no longer sure what you are looking at. However, **Number of Sales for Each Year Multiplied by the Average Price of Service per Year for the Years 2010, 2011, 2012, 2013, 2014, 2015, 2016, 2017, 2018, 2019, and 2020 as Aggregated from Quarterly Reports** is a painfully long title that ends up being much more confusing. Finally, **Cost of Living per Region** has nothing to do with what is actually on the chart, so while clear and concise, it is not accurate. Always make sure the title, labels, and legends reflect what is in the table as well as being clear and concise. Keep it clean, keep it simple.

Color theory

Color theory is a set of rules that identifies which colors complement one another and which colors clash and actually cause stress to the eyes. Again, if your company has a brand that includes a color scheme, that is what you should use, but if you have to make your own, then this is something you should learn at least at a basic level. Trying to show and explain more about it in this book is pointless because it will be printed in black and white. Besides, for the exam, all you need to know is to apply company colors first, and if there are no company colors, then at least make sure they are easy to read and not painful to the eyes.

If you want to know more about color theory, there are a ton of resources, free and otherwise, that will give you a basic rundown. A fun one is `www.canva.com/colors/color-wheel/` because it is interactive and lets you see how the different schemes work.

Summary

This chapter has covered a wide range of topics because there are a lot of things to consider when making a report. First, we went over the development process step by step. Then, we looked into business requirements and how they impact your report, as well as considerations specific to dashboards. Next, we talked about the different elements that have to be included or updated in a report. Then we went over everything you have to think about while delivering a report. Finally, we went over basic design elements that will make your report easy to understand and look more professional.

In the next chapter, we will actually go over popular visualizations, which should be a fun chapter. See you there!

Practice questions

Let's try to practice the material in this chapter with a few example questions.

Questions

1. What is the first step in the development process for a dashboard?

    A. Get approval

    B. Make a plan

    C. Create a dashboard

    D. Deploy the dashboard

2. At a dairy farm, a low-level manager is responsible for a certain number of cows. They would like a report a couple of times a month on how much milk each cow is producing to see whether any of them require special care. What is an appropriate audience for this report?

    A. Only the manager

    B. Every manager

    C. The manager and the farmhands

    D. Everyone in the company needs this report

3. In the scenario from the previous question, what is the most appropriate frequency of the report?

    A. Daily

    B. Weekly

    C. Monthly

    D. Yearly

4. The day after releasing your milk report, you find a major mistake, and have to rerun the report to create a new version, but can use the data from the original report. Which of the following report elements will need to be updated?

    A. FAQs

    B. The version number

    C. The version number, the report run date, and the data refresh date

    D. The version number and the report run date

5. You decide to create a static dashboard for the milk production report, and you want it to automatically refresh and send out invitations every Sunday night. Which delivery consideration should you use?

    A. Filtering

    B. Scheduled delivery

    C. Subscription

    D. Dashboard optimization

6. While designing the dashboard for milk production, you have to pick a color scheme for your charts. What colors are most appropriate?

    A. High-contrast colors

    B. Low-contrast colors

    C. Moderate-contrast colors

    D. Company brand colors

Answers

Now, we will briefly go over the answers to the questions. If you got one wrong, make sure to review the topic in this chapter before continuing:

1. The answer is: Make a plan

    The first step of the development process is to make a plan. You cannot get your plan approved, create the dashboard, or deliver it without planning first it.

2. The answer is: Only the manager

    The information in the report is only applicable to this one person. The manager might make decisions based on this report that will impact those above or below them, but they are the only one who is directly interested in the production rates of these cows.

3. The answer is: Weekly

    The question specifically states that they want the reports a couple of times a month, and this is the only option that fits that requirement.

4. The answer is: The version number and the report run date

    The report was rerun on a different day, generating a new version of the same report. This requires a new version number and a new report run date. The question states that you used the same data, so the data refresh date does not change, and there is nothing in the question about the FAQs.

5. The answer is: Subscription

    Subscription is the only option that automatically delivers an updated version of the report at the same time every week.

6. The answer is: Company brand colors

    Remember that the company brand takes priority over every design decision. The amount of color contrast does not matter at all if the company has a color scheme as part of its brand. Always use branding guidelines if they exist.

13
Common Visualizations

This chapter is all about common visualizations, which are important tools for every data analyst. As we discussed in the previous chapter, data analysts need to be able to communicate results to other people. They say a picture is worth a thousand words and it is true that data visualization is often a more effective form of communication than just reading out the numbers of your results. Almost every report you create will have at least one visualization, often more. There are all kinds of ways to depict your data, but in this chapter, we will cover the most common types, which will also be the types you need to know for the exam.

Here, we will discuss infographics and word clouds. Then, we will move on to bar charts and bar chart variations. Next, we will cover charts with lines, circles, and dots. Finally, we will wrap it up with mapping visualizations. In the exam, you will be asked both what kind of chart a given illustration depicts and what kind of chart is most appropriate to use in a given circumstance. There is a lot to cover, but visualizations are fun, so let's get started!

In this chapter, we're going to cover the following topics:

- Understanding infographics and word clouds
- Comprehending bar charts
- Charting lines, circles, and dots
- Understanding heat maps, tree maps, and geographic maps

Understanding infographics and word clouds

Infographics and word clouds are useful tools. There are analysts who specialize in infographics specifically, but we will talk more about that in a moment. First, while these can show data or numbers, infographics and word clouds are more often used to communicate more general concepts. They are visuals that express ideas or generalizations more than hard facts. That said, when you need to explain the flow of data through your ETL pipeline, a pie chart isn't going to cut it and maybe a flowchart won't be quite right either, so they are still an important part of the data analyst's toolkit.

Infographics

As we said before, some analysts specialize in infographics because sometimes you need to explain broader concepts or there just is no chart for the specific thing you are trying to communicate. Infographics are particularly useful when there is a concept you need your audience to understand for the rest of the report to make sense, or if you want to communicate the concepts in a pretty way in reports released to the general public. In fact, infographics can actually contain other types of visualizations.

There are books and guides on how to make effective infographics, and there are even specialized tools. If you are interested, you can check out www.visme.co/make-infographics/ for a free tool.

Let's look at an example in *Figure 13.1*:

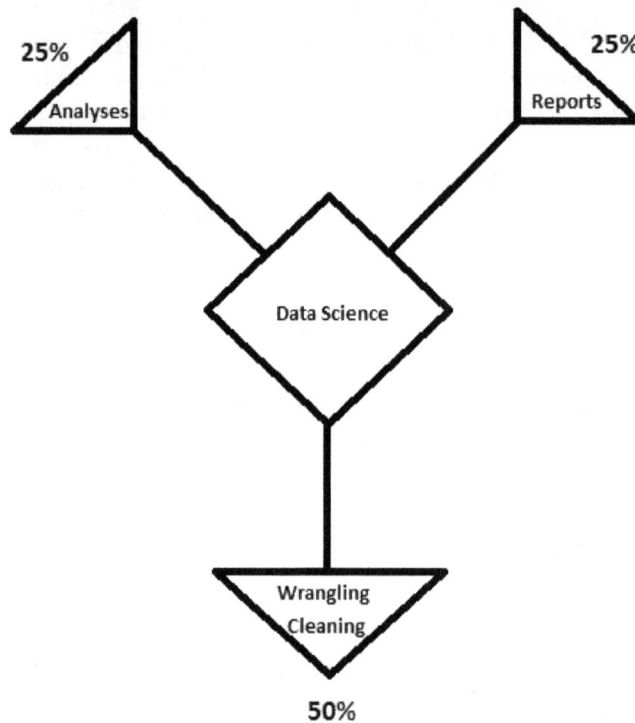

Figure 13.1 – Infographic

Here, you can see a clear and simple take-home message, and that is why infographics are effective. They should speak for themselves so that you don't need to sit down and explain anything. With little to no background, any pedestrian should be able to get a general idea. These are great for when you have a broader concept that you need to get across without a lot of space to do so.

Word clouds

Word clouds are a bunch of words of different shapes and sizes randomly spread out in a cloud-like shape and can be used in a few different ways. Technically, they are supposed to be calculated and generated based on data, as with the rest of your charts, but that isn't always the case. You can think of word clouds as a simple form of natural language processing; they take text and break it down into words, and then word sizes and placements are based on the frequency of the word and how it is used. To put it another way, the more often the word comes up in the text, the bigger and more prominent it is in the word cloud. In this way, you can see what a piece of the text focuses on. Sometimes, these are generated artificially – *gasp* – just to express the main idea and the supporting ideas behind it.

Okay, so why do this? Well, it is a popular visualization for natural language processing analyses. Word clouds are more advanced than expressing text information as a strict frequency count; it figures out which words are important and how important they are. For example, if you were to put the text in this book into a word cloud, it might look something like *Figure 12.2*:

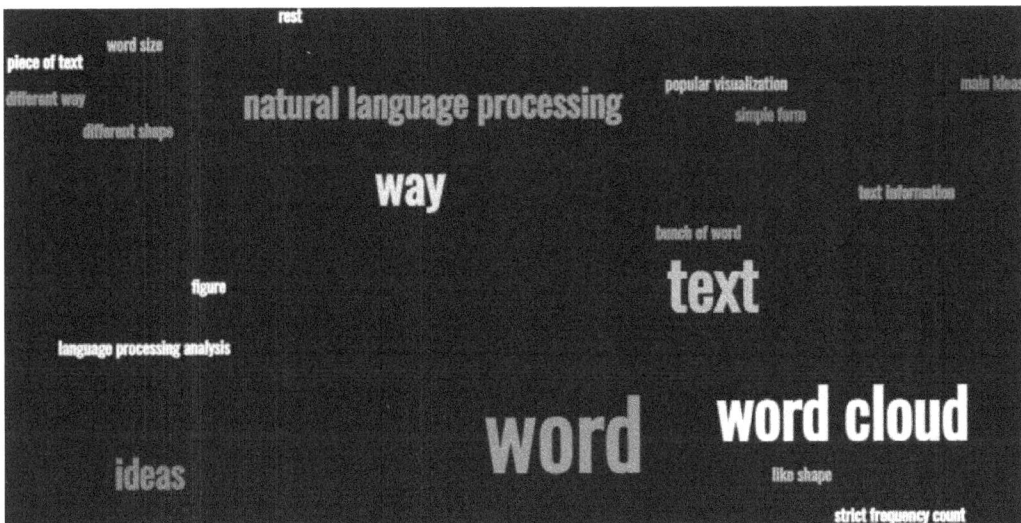

Figure 13.2 – Word cloud

There are several ways to make these, but, if you just want to play around a little bit, you can try a free tool, such as www.monkeylearn.com/word-cloud/. Try it out, and if you are trying to visualize a text analysis, think of word clouds first. Let's move on to charts with bars!

Comprehending bar charts

One of the most iconic charts is the bar chart. Everyone has seen one and they are sometimes taken to symbolize the general concept of a chart. What you may not know is that there are actually several different kinds of bar charts that are used for different situations. Here are a few of the most common ones.

Bar charts

The classic bar chart is the simplest form of bar chart, but also one of the most useful. Bar charts are used to compare one quantitative variable to a qualitative variable. In other words, you are looking at a number that is broken down into categories. We can see what this looks like in *Figure 13.3*:

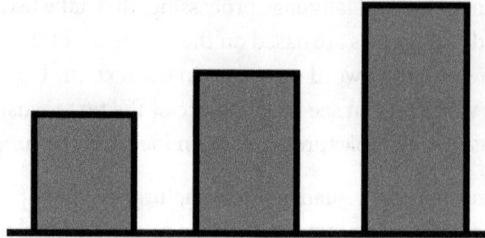

Figure 13.3 – Bar chart

Each bar represents a group, and the size of the bar represents the number value for that group. This is the bread and butter of data visualizations and is used in pretty much every field.

Stacked charts

Stacked bar charts are a little more complicated. Here, you are looking at two different qualitative variables (or two levels of the same qualitative variable), and one quantitative variable. Effectively, you are trying to cram a second qualitative variable into a regular bar chart, so you are comparing two sets of categories using one number variable. You can see this in *Figure 13.4*:

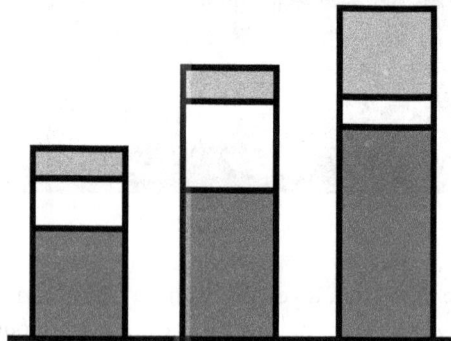

Figure 13.4 – Stacked bar chart

Now, as you can see, it looks like a bar chart, but now each bar is in multiple colors. Each bar represents a different group from the first qualitative variable, each color represents a different group from the second qualitative variable, and the size of the bar and the colored bands within the bar represent the numerical value of the quantitative variable for each group. For example, you have three different people who make paper airplanes. All of them make the same three designs. With a stacked bar chart, you can keep track of how many paper airplanes each person made, and look at how many of each design each person made all in one chart.

Histograms

Histograms are very similar to bar charts, and, in some cases, there is little to no difference. A histogram is a bar chart that compares one quantitative variable to a scale. Yes, a scale is an ordinal variable, which is a type of categorical variable, so how is this different from any other bar chart? Well, the specific type of categorical variable matters. These are usually used to compare two quantitative variables, but one of those variables is turned into a scale. You break one of your quantitative variables into even chunks, sometimes called buckets or bins, and each chunk becomes a group or bar on the chart.

This is why, when visualized, histograms often have the bars touching, because one starts exactly where the last one ends. What is the point of this? There are lots of uses for histograms, but the most common one is to visualize distributions. Let's see what this looks like in *Figure 13.5:*

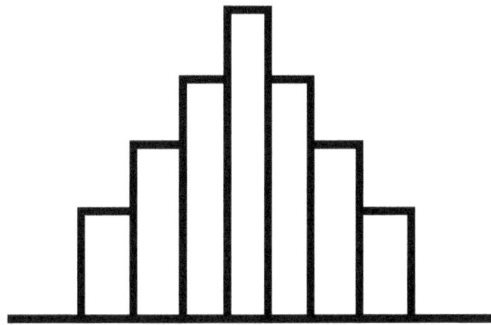

Figure 13.5 – Histogram

Here, each bar represents a different group of the scale, and the size of the bars represents the value of your quantitative variable. Imagine the classic distribution of grades on a test. The grade is a scale that breaks down into groups, each having its own bar, and the size of the bar is how many students got grades that fell into that group. That is how histograms work.

Waterfall charts

Waterfall charts are a special kind of bar chart that focuses on the differences between the bars. These are popular for project management. They compare one quantitative variable to one qualitative variable. The qualitative variable is often different points in time but can be any group. These are unique in that the middle bars only show the difference between that group and the group on the left. For example, you can use waterfall charts to show progress from week to week, and it highlights how you did each week to how you did the previous week. Are you getting better, worse, or are you staying steady? We can see what one looks like in *Figure 13.6:*

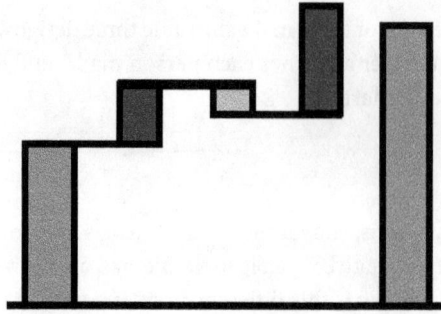

Figure 13.6 – Waterfall chart

These can be a little tricky to read until you get used to them. The bar on the far left is the starting point, the bar on the far right is the final or summary point, and each partial bar in the middle shows the change from the previous bar. Often, positive changes are a different color from negative changes, and the starting and ending points are a third color. This way, you can see at a glance which weeks showed improvement and which didn't.

That wraps things up for bar charts; next, we will go over some of the other shapes, such as lines, circles, and dots.

Charting lines, circles, and dots

There are several types of charts that do not contain bars. You have probably heard of some of these, but perhaps not all of them. We will briefly go over common charts that contain lines, circles, and dots and how they are used!

Line charts

The line chart is another iconic chart and is used pretty heavily in certain fields. A line chart tracks changes in a quantitative variable over time. Time is always on the *x*axis and your number variable is on the *y*axis. These are usually used to look at general trends – is this variable increasing over time, decreasing over time, or staying roughly the same? Alternatively, it can be used to see whether a specific point is breaking a historical trend. Let's see what a line chart looks like in *Figure 13.7*:

Figure 13.7 – Line chart

Effectively, a single point is made for your quantitative variable for every value on the *x*axis, which can be any unit of time. Then, the dots are connected to create a line, showing how the quantitative variable changes over time.

Pareto charts

Pareto charts are the one type of chart in this chapter that are not on the exam. They are covered here because they are annoying and pervasive in the field and can come up in interview questions. Personally, I do not like Pareto charts. They are messy and needlessly complicate matters. That said, they are good to know about.

A Pareto chart is a mash-up of a bar chart and a line chart. As you recall, a bar chart is one quantitative variable and one qualitative variable, and that doesn't change in the Pareto chart. The line chart, however, is slightly different. The line chart uses the same quantitative variable as the bar chart *and* the same qualitative variable, so it does not have to use time on the *x*axis. This looks something like *Figure 13.8*:

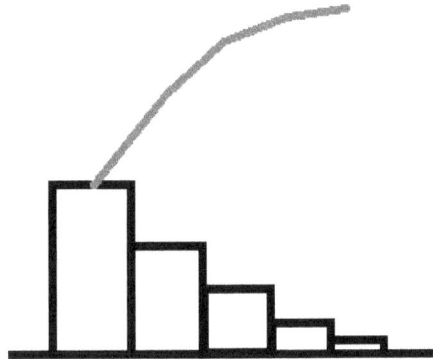

Figure 13.8 – Pareto chart

Effectively, you make a bar chart first and sort it so that the bar with the largest value of your quantitative variable is on the far left and the other bars are in descending order moving left to right. Then, you have a line chart that tracks the cumulative sum of the quantitative variable, which will progressively go up with each group as the value of that group is added to it. This gives you a bar chart that goes down from left to right and a line chart that goes up from left to right.

Pie charts

Pie charts are another classic chart that everyone should know. These show a qualitative variable broken into the percentages of each group to make a whole. For instance, if you are looking at test results and the only options are pass or fail, the percentage of people who passed would be one color on the circle and the percentage of people who failed would be another color. The pie chart should account for all possible outcomes, so the circle should always be full. Let's see what this looks like in *Figure 13.9*:

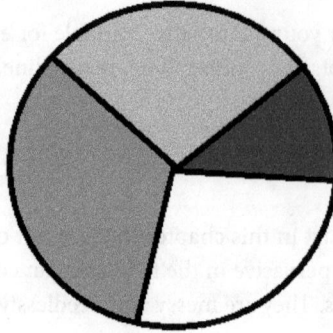

Figure 13.9 – Pie chart

These charts are pretty straightforward and all you need is one qualitative variable. This makes them great for things such as demographic data. How many of your customers are in each age bracket? Of each race? Gender? Education level? This can help you narrow down who your customers are and apply more targeted advertising.

Scatter plots

Scatter plots are very useful. We have mentioned them multiple times in this book already because they are heavily used to see whether there is a relationship between two variables and even whether that relationship is linear. Again, this compares two quantitative variables. We have seen a few of these already, but let's look again at *Figure 13.10*:

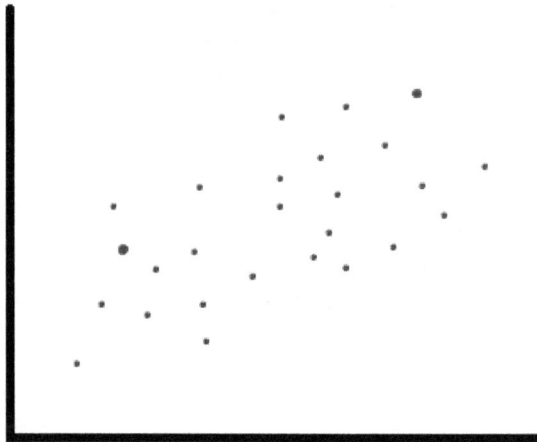

Figure 13.10 – Scatter plot

Again, these two quantitative variables need to be paired. That means that every observation has to have a value for both of them. Each dot represents a single observation, and its location depends on its values for the two variables.

Bubble charts

Bubble charts are kind of fun. They are a variation on a scatter plot. The difference is that instead of two quantitative variables, a bubble chart shows three quantitative variables! Again, all three of the variables need to be paired and each bubble represents a single observation. Let's see what this looks like in *Figure 13.11*:

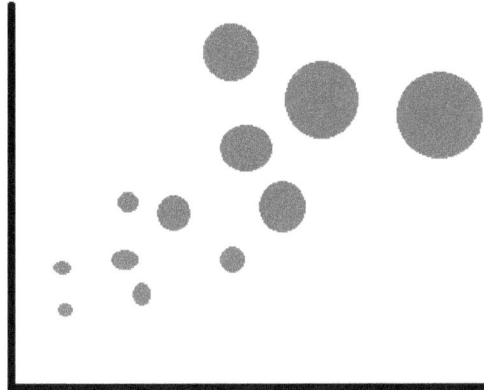

Figure 13.11 – Bubble chart

This is like a scatter plot, with one quantitative variable on the *x*axis and one quantitative variable on the *y*axis. The third quantitative variable is represented by the size of the dots. The bigger the dot, the higher the value. This can still show relationships between the three variables; maybe the dots get bigger or smaller as they continue in a given direction. That said, they are harder to interpret, and you cannot show as many observations at a time, or they will all blend together and become meaningless. It is up to you whether you use a bubble chart or multiple scatter plots, but make sure you know about it for the exam.

Next, let's get into mapping visualizations!

Understanding heat maps, tree maps, and geographic maps

The final category of visualizations you need to know for the exam is the maps: heat, tree, and geographic. These may or may not be what you would think of when you hear the word map. A map is a visualization that depicts the relationship between different elements based on some variable. A traditional map, a geographic map, shows the distance and direction between separate points so that you can see how to get from one area to another. However, not all maps denote locations in space like a geographic map does. Let's jump right in and you will see for yourself.

Heat maps

Heat maps are interesting and often colorful. A heat map compares two qualitative variables, almost always scales, and one quantitative variable. These are designed to see whether there is a relationship between the two scales and that relationship is reflected in the changes in the quantitative variable. It will be easier to see one, so let's look at *Figure 13.12*:

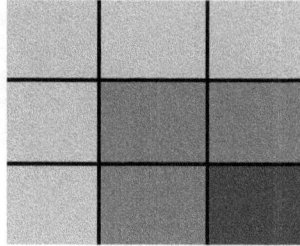

Figure 13.12 – Heat map

Okay. The *x*axis and the *y*axis represent two scales. The color of the cells where they intersect on the grid represents the quantitative variable. Heat maps require a legend, so the person reading them knows what color corresponds to what values. Do the values go up with darker colors or down?

This is kind of confusing, so let's go over an example. You work for a company that has stores all over the country. You want to see whether the size of the store and the size of the city in which they are located impact how much money they make. You create a scale for store sizes: **Small**, **Medium**, and **Large**. Additionally, you make a scale for city sizes: **Small**, **Medium**, and **Large**. The color of a cell represents the average income of stores that meet those criteria. Let's see this in *Figure 13.13*:

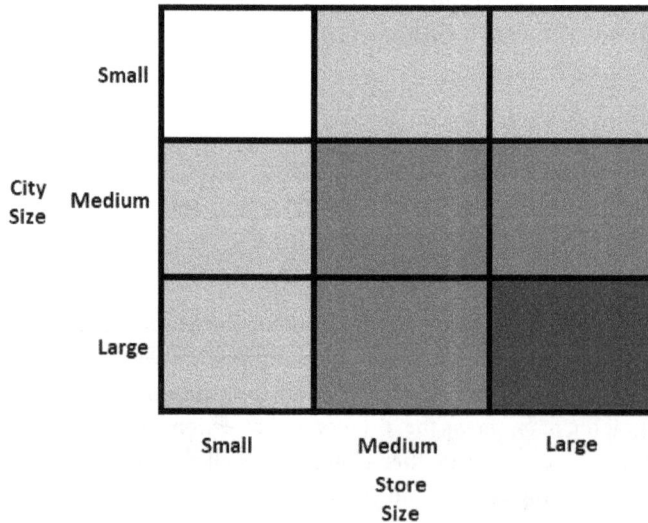

Figure 13.13 – City versus store size heat map

This is pretty straightforward, and we can see that there is a relationship between city size and store size. Large stores in large cities make the most money and small stores in small cities make the least money. However, both qualitative variables play a role, because neither city size nor store size alone will yield the best results. Neither a large store in a small city nor a small store in a large city will make very much.

These are used in some fields more than others, and it really depends on the type of data with which you are working. You either need scales and the desire to compare them, or the ability to make scales. Heat maps are often used to emphasize contingency tables.

Tree maps

Tree maps are a little weird and often need some explaining, because they are used to depict complicated relationships, and end up looking convoluted themselves. Here, you are comparing a nested or hierarchal qualitative variable with a quantitative variable. Then, the colors of the boxes represent groups or levels. The problem is that sometimes people decide that instead of using a nested or hierarchal qualitative variable, they want to sneak in a second quantitative variable for color in a sort of tree map/heat map hybrid. The good thing about tree maps is that your qualitative variable does not need to be a scale. Let's look at a tree map in *Figure 13.14*:

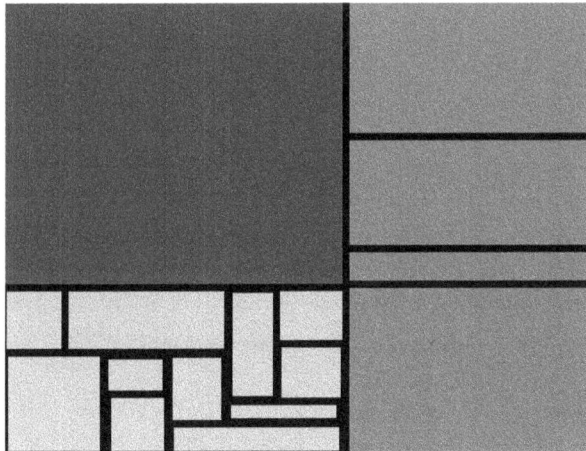

Figure 13.14 – Tree map

Each box represents a group in your qualitative variable. The size of each box represents your quantitative variable. The color of each box either shows another level of your qualitative variable OR an additional quantitative variable.

Sounds confusing? Let's just consider an example. You work at a company with different divisions and each division sells its own products, and you want to know which products are selling the best, especially in regard to which divisions are selling them. Let's see this in *Figure 13.15*:

Product Sales By Department

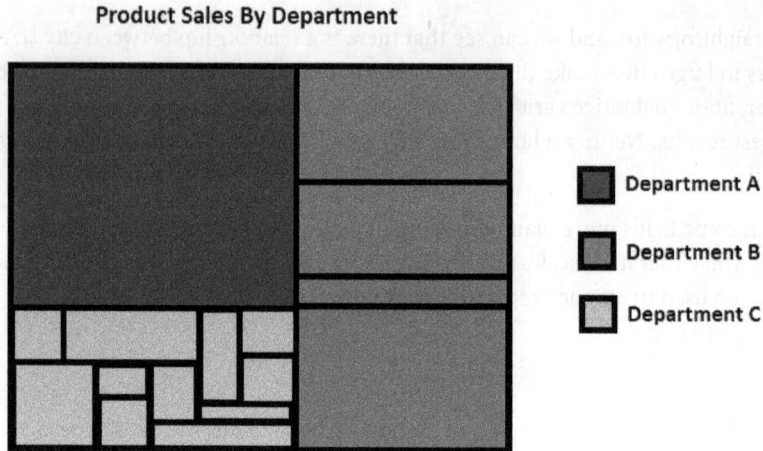

Figure 13.15 – Product sales by department tree map

Each box represents a product, the color of the box describes the division – **Department A**, **Department B**, or **Department C** – that made that product, and the size of the box is how many of that product have been sold. Here, you may be able to see general trends in the data. For example, **Department C** has more products, but none of them sell very well, while **Department A** has only one product, but it sells very well. **Department B** falls somewhere in the middle both in terms of the number of products and how well they sell.

Geographic maps

Geographic maps look like, well, maps. There are a few different approaches to these, but, in general, they compare physical locations (a qualitative variable), to a quantitative variable. These are pretty self-explanatory, so let's jump right in and see what one looks like in *Figure 13.16*:

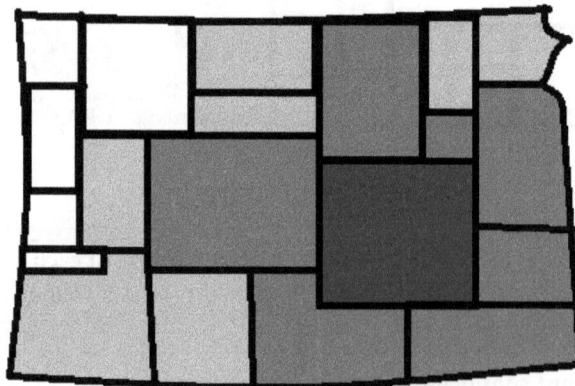

Figure 13.16 – Geographic map

Often, it looks like a map broken down into chunks and the color of the chunks represents your quantitative variable. Sometimes, it will look like a map with dots and the size or color of the dots will represent your quantitative variable. These are not very standardized, but they are pretty easy to interpret. If you ever see a map of a physical location, it will be a geographic map.

Summary

There was a lot of information covered in this chapter, but these are all useful tools. You learned about 14 different kinds of charts, including infographics, word clouds, bar charts, stacked bar charts, histograms, waterfall charts, line charts, Pareto charts, pie charts, scatter plots, bubble charts, heat maps, tree maps, and geographic maps. Whew! Good job! Make sure you know them by sight and by when they are appropriate to use. Pay attention to what types of variables they use and how many. This wraps up the chapters on reporting and visualizations. Next, we will dive right into the higher level of data governance!

Practice questions

Let's test our knowledge of the material in this chapter with a few example questions.

Questions

1. The following data visualization is representative of which type?

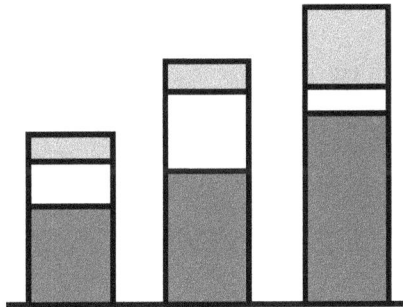

    A. Bar chart

    B. Stacked bar chart

    C. Histogram

    D. Pie chart

2. If you wanted to represent a single quantitative variable over time, which data visualization type would be most appropriate?

   A. Scatter plot

   B. Bar chart

   C. Heat map

   D. Line chart

3. The following data visualization is representative of which type?

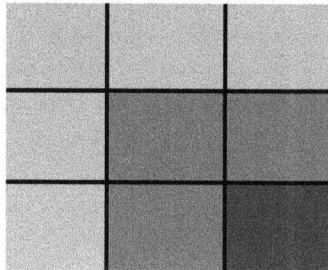

   A. Heat map

   B. Tree map

   C. Geographic map

   D. Bar chart

4. If you wanted to track a quantitative variable over a country, which type of data visualization would be most appropriate?

   A. Bubble chart

   B. Geographic map

   C. Heat map

   D. Tree map

5. The following data visualization is representative of which type?

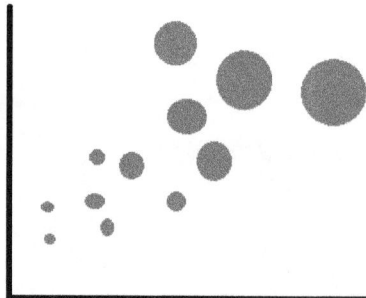

A. Scatter plot

B. Dot plot

C. Bubble chart

D. Pie chart

6. A project manager wants a chart to track progress throughout a project, and they want to know, specifically, how each point differs from the previous point. Which data visualization is most appropriate?

A. Waterfall chart

B. Line chart

C. Bar chart

D. Stacked bar chart

7. If you wanted to display a qualitative variable with two different levels, and a qualitative variable on the same chart, which type of visualization would be most appropriate?

A. Waterfall chart

B. Line chart

C. Bar chart

D. Stacked bar chart

Answers

Now, we will briefly go over the answers to the questions. If you got one wrong, make sure to review the topic in this chapter before continuing:

1. The answer is: Stacked bar chart

    The displayed chart is a stacked bar chart. You can tell because the bars are divided with different colors representing different groups.

2. The answer is: Line chart

    While a few of these technically could be used to track a quantitative variable over time, the line chart is specifically designed for it, making it the most appropriate.

3. The answer is: Heat map

    Heat maps show regular grids representing ordinal variables, usually with a quantitative variable representing the color.

4. The answer is: Geographic map

    Since we want to look at data over physical space, in this case, a country, a geographic map would be most appropriate.

5. The answer is: Bubble chart

    Here, we see what looks like a scatter plot, but the dots are different sizes, representing a third quantitative variable.

6. The answer is: Waterfall chart

    Again, there is more than one type of chart that can display this information, but waterfall charts are designed for this kind of visualization.

7. The answer is: Stacked bar chart

    Remember that stacked bar charts represent two qualitative variables OR two levels of the same qualitative variable and a quantitative variable.

14

Data Governance

In this chapter, you will learn about data governance. Data governance is kind of a broad topic that is applied on multiple levels, but at its heart, it is about the policies meant to protect people and the integrity of the data. It can also be used to keep data clean, accessible, and easy to use. These policies can be international standards, national laws, local laws, industry regulations, company contracts, or even personal rules. Data analysts almost never have a say in data governance, and many ignore it completely assuming that if they accidentally break a rule, it isn't their fault because their supervisor didn't tell them. This is a mistake. When you are working with data, it is your personal responsibility to make sure you know what the rules are and how they apply to you. Ignoring data governance can hurt the people whose data you are using and have legal repercussions for you personally.

The rules change wildly based on where you live, what industry you are in, what company you work for, and even what specific types of data you use. This chapter cannot cover every rule for every person, but CompTIA does want to make sure you are aware of what types of policies exist and what you should look for when preparing yourself. First, we will cover data security and how it applies to data governance. Then, we will go over data use agreements and their important sections. Next, we will talk about the different data classifications. Finally, we will cover entity relationship requirements.

In this chapter, we're going to cover the following main topics:

- Understanding data security
- Knowing use requirements
- Understanding data classifications
- Handling entity relationship requirements

Understanding data security

In this section, we will discuss data security. Data is the most important resource for a data analyst. Without data, we would be out of a job. Not only is it important to have data, but it is important to make sure the data maintains its integrity. **Data integrity** is basically how valid, or accurate, the data is, and maintaining data integrity involves work on several different levels. For example, using the data to calculate a new variable does not impact the original data at all; the data itself hasn't been touched and there is no impact on data integrity. However, changing or manipulating the data to show trends that are not there does impact data integrity. Incorrect use of the data means the data becomes useless because the sample will no longer reflect the population. Through malice or mistake, if anyone can access the data from anywhere at any time, then the data is at serious risk. Here, we will discuss how to keep the wrong people from messing with your data and damaging its integrity with data security.

Data security serves many roles, all of which are important, and some are legally required. Not only does it help maintain data integrity, but it also keeps other companies from using the information you worked so hard to get against you. Also, it is legally required for working with protected data classifications, which we will discuss later in this chapter in the *Understanding data classifications* section.

Often, it is not the job of the data analyst to implement data security protocols (that honor usually goes to a cyber security specialist), but it is the job of the data analyst to understand and follow those protocols. Let's go over the most common types of data security, which will be in the exam.

Access requirements

Access requirements are the most basic level of data security. These simply limit who has permission to access the data. Limiting access to the data to a few specialized professionals makes it much more likely that it will stay safe. However, most companies do not decide who gets access on a person-to-person basis, because it would just take too long. Instead, there are the following two main ways in which access is granted:

- Role-based
- User group-based

Role-based access focuses on the role a person plays in a company. For example, everyone in a data analyst role may have access to a general dataset for the entire company, no matter which department they are in or what they do on a daily basis.

User group-based access focuses on the specific group to whom the data pertains. For example, only people in the marketing department can access the marketing data, or only the people on the project team can access the project data.

Often, a company actually uses a combination of these methods, based on the situation. However, in the exam, you will most likely be given a scenario and asked which method is most appropriate. Just ask yourself whether this data pertains to a job title across the company, or to a specific group of employees.

When setting access requirements, you also need to consider **Data Use Agreements**, which are contracts that state specifically how data can be used, processed, deleted, and maintained. We will discuss this further in the *Knowing use requirements* section. Until then, simply realize that these contracts can limit who you can let see the data. If you want to share the data with anyone not listed in the Data Use

Agreement, you need to get their express approval with a **release approval**, which is another contract that allows the data to be shared with a person, group, or entity. You may have encountered this when changing your doctor. To have your medical records transferred to the new doctor's office, you will need to sign a release approval document.

Security requirements

Besides access requirements, there are a few other general types of data security that you will run into. You don't need to be an expert in cyber security, but you at least need to be familiar with these general concepts:

- Data encryption

- Data transmission

- De-identification/masking of data

Let's start from the top. **Data encryption** uses algorithms to translate data from plaintext to cyphertext. In other words, it makes the data impossible to read. The only way to use the data is to translate it back into plaintext, which requires a key or the original algorithm that encrypted it in the first place. Basically, this means that even if someone who shouldn't have your data gets their hands on it, they can't use it.

> **Important note**
>
> There are different laws about what types of data need to be encrypted, when they need to be encrypted, and even how complicated the encryption needs to be. Some areas even have laws about whether or not you even can encrypt your data. These vary from country to country around the world. If you are unsure about the encryption requirements of your area, `https://www.gp-digital.org/world-map-of-encryption/` is a great resource.

Data transmission is how the data is moved from one storage location to another. Data in transit can be vulnerable, so it is important to protect it. For example, downloading legally protected information to an unsecured personal laptop through an unsecured public connection is probably a bad idea, and likely illegal. The data could be stolen, altered, or corrupted as it is being downloaded. You don't need to be an expert in network security, but it is important that you understand that data is more open to danger while it is being transmitted. Make sure you use the approved secure connections provided by your company, and never take your data home without express permission.

De-identification/masking of data describes the process of removing any personal or sensitive information from your data so that it is legally safe to report. No one wants a bank to give out all of their financial information; that is just begging to have their identities stolen. However, a bank can remove any information that can be linked back to any one person to release a report about how well they did last year, to inspire people to have confidence in them and attract new customers.

Data security is something that you will probably never be in charge of but will interact with daily. Knowing these general concepts helps you understand why data security is important and why you need to follow security protocols.

Some data security requirements are set not by a government official, but by a contract with the person from whom you are collecting data. In the next section, we will talk about use agreements.

Knowing use requirements

The use requirements are set by the Data Use Agreement. If you have ever installed a new piece of software and before you could use it, you needed to agree to a Terms of Use policy, then you have been on the receiving end of one of these policies. It states what data they can collect, what they can do with it, how they will handle it, when they will delete, and whether they can retain it.

The Data Use Agreement is a legal document, and it is unlikely you will ever need to write one as a data analyst. However, you need to know what rules are set in the Data Use Agreement document used by your company. This clearly sets boundaries on how the data can be handled. Deviating from the Data Use Agreement is illegal and will have consequences for the company and for you personally. This may sound cynical, but don't just trust that what another analyst says should be fine. Look at the agreement and know for sure. As the one using the data, it is *your* responsibility to make sure you are using it correctly.

Now that that is out of the way, let's look at different parts of the Data Use Agreement.

Acceptable use policy

The **acceptable use policy** is, as you would imagine, an important part of the Data Use Agreement. This section has these subsections: preamble, definitions, policy statement, acceptable uses, unacceptable uses, and violations/sanctions. Luckily, you don't need to memorize every part. You just need to have a general idea that the acceptable use policy says how they can use the data, how they can't use the data, and what happens if they break the policy.

Data processing

Data processing details how the data can be processed. Surprising, right? This is less about how you will use the data and more about how you will treat it during interim steps, such as transfers to protect the security and rights of the people from whom you are collecting the data. This can also include statements about training on how to handle the data, whether or not a data protection officer is assigned to the data, or whether a data protection impact assessment will be performed. The data processing section of the document tells you that, in general, the data will be handled in such a way as to protect the person who gave it.

There are all kinds of laws and regulations around the world that detail how this section should be handled, but the gold standard in the field is probably the European Union **General Data Protection**

Regulation (**GDPR**). If you are interested in more information on GDPR, you can look it up at `https://gdpr-info.eu/`. This is a great resource for how to treat data.

For the exam, you need to have a general idea that this section deals with: how the data is treated during transfers and data breaches, and what steps you are taking to keep private information private. For your job, this section details some of the security protocols, such as masking data or using secure networks for data transfers. Before we move to the next section, we need to focus on an important element found in the data processing section: data breaches.

Data breaches

The details for how to handle data breaches are outlined in the *Data processing* section, but they are so important that they get special attention in the exam. A **data breach** happens when your data security is compromised and someone has access to your data when they shouldn't. If you even suspect a data breach, follow these steps:

1. Report the breach.
2. Secure operations.
3. Fix vulnerabilities.
4. Notify the impacted parties.

Basically, you let your boss know there might have been a breach, they make sure things are safe, they make sure it doesn't happen again, and then you must let anyone whose data *might* have been accessed know about the breach. The important steps are the first and the last: report any breach and notify the impacted parties. Staying silent is the worst thing you can do.

Data deletion

Aptly named, the **data deletion** section explains how and when the data will be deleted. While some companies try to hold onto your private information forever, there are several good reasons for deleting it:

- Consent is withdrawn
- Illegal means were used to collect or process the data
- Legal obligations
- The person who gave the data was a minor or not of sound mind
- Data is no longer needed
- The data retention period ended

Most of these are legal reasons, but there are also practical reasons. Keeping data you don't need takes up space, and storage costs money. It also slows down your system, makes things more complicated, and after a certain point, you need to hire more people to manage your greater storage. More data is not always better, despite popular opinion. It is best practice to delete data that has been used for its intended purpose and is no longer required.

Data retention

Data retention usually covers how long they will keep the data and how it will be stored. This also covers security protocols that are based on the stored data, such as who gets access to it. Sometimes, this includes a data retention date, which says that after a certain point, it will be deleted. You just need to know that this section details how and for how long they will store the data.

That wraps up the Data Use Agreement. It is important to know what your company's version of this document says because it is a binding contract between the person and the company, and you are acting as a representative of the company. This means that you are a party in the contract and responsible for it. Doesn't it make sense to at least read it?

That's enough about contracts; let's talk about other laws that apply to certain types of data by default and can get you in trouble. Specifically, let's talk about protected data classifications.

Understanding data classifications

Certain types of data are legally protected. No one cares if you create a dataset of measurements of caterpillars that you find in your local park; you can use that data with reckless abandon! However, if you are using information that has the potential to hurt someone, there are rules about how to treat it. There are three main data classifications that fall into this category.

Personally identifiable information

As the name indicates, **personally identifiable information** (**PII**) is a classification of data that includes anything that can be used to identify a specific person. There was a time when this meant that you just couldn't give out data with peoples' names on it, but there are now many ways to find how who a person is based on some seemingly obscure information. This includes, but is not limited to, the following:

- Name
- Physical address
- Email address
- IP address
- Social Security number
- Phone number
- License number
- Passport number

- Login ID

- Social media ID

- Social media posts

- Date of birth

- Digital images

- Geolocation

- Biometric data

- Behavioral data

Any data that can even theoretically be used to track down or identify a specific person is considered PII. It does not mean that you cannot have PII in your dataset, sometimes you need it, but you must keep it secure and *never* report it. Some of this will be covered in the Data Use Agreement, but there are also federal and even local laws regarding how this data can be treated. Make sure you know which laws apply to you.

Personal health information

Personal health information (**PHI**) is identifiable information that relates to the past, present, or future health of a person. It is similar to PII and is sometimes considered a type of PII, but it gets its own data classification because there are laws around it specifically. PHI covers any and all medical records, including dental records. Any information about your health is considered PHI and is protected. There are laws protecting health information all around the world, but the most heavily referenced is the **Health Insurance Portability and Accountability Act** (**HIPAA**). You can learn more about it at `https://www.hhs.gov/hipaa/index.html`, which is a good resource. However, wherever you are in the world, check to see the legal requirements for how to handle health-related information.

Payment Card Industry

The **Payment Card Industry** (**PCI**) is all about financial information, specifically relating to credit and debit cards. Technically, this means that a company is compliant with the Payment Card Industry Security Standards Council. You can learn all about that at `https://www.pcisecuritystandards.org/`. This also focuses on protecting **personally identifiable financial information** (**PIFI**). Again, some people consider this a type of PII, but it gets its own classification because it has laws focused on it specifically. This can get kind of confusing, but for the exam, all you need to know is that any credit card information is considered PCI and is protected.

That's all for data classifications. Let's move on to entity relationships.

Handling entity relationship requirements

Entity relationship requirements are sometimes laws, but more often, rules about how pieces of data can relate to one another. An entity is a table, model, or data object. Entity relationships are how these data objects connect to each other. You can see how this happens in an entity relationship diagram, as shown in *Figure 14.1*:

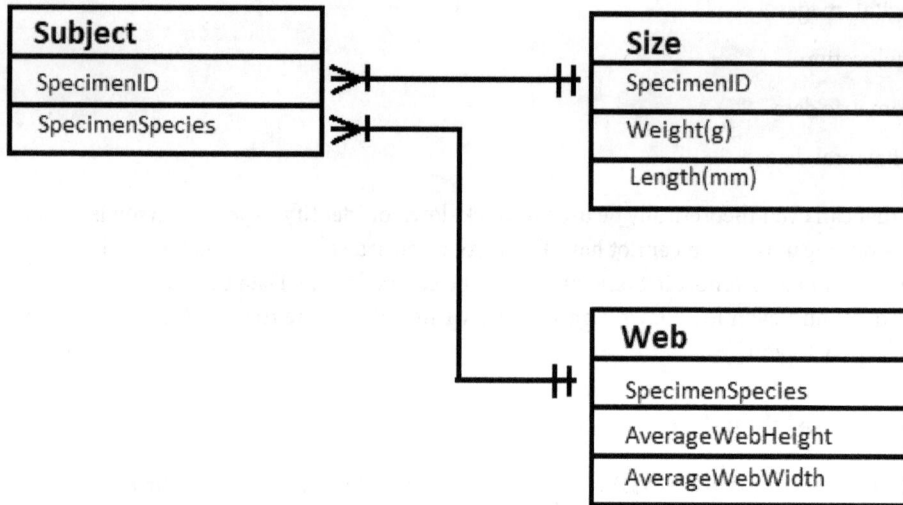

Figure 14.1 – Entity relationship diagram

This looks a lot like a flowchart and shows not only that the tables are connected, but often the specific variables that connect them, and the markings on the line can denote what kind of relationship these variables have. Again, you will need to look up what the entity relationship requirements are for you, but there are a few main types of requirements:

- Record link restrictions

- Data constraints

- Cardinality

Let's go ahead and look at these in a little more detail.

Record linkage, or data linkage, is, as you would guess, linking different pieces of data of the same record. **Record link restrictions** mean that those two pieces of data must never be linked together. This seems odd, but imagine you had two pieces of information for one person. Individually, they are safe and there are no problems, but when they are together, they are suddenly PII and considered protected data. Alternatively, imagine a company that had every single piece of information required to steal your identity and empty out all of your bank accounts. You would certainly hope they wouldn't keep it all in one easily accessed place.

Data constraints are rules designed to protect data integrity. There are many ways data constraints can be applied, but the most common by far is a set of rules that specify what types of data can be entered into the database, the condition of the data, the format of the data, and even how or when it is entered. Data constraints make sure only the highest quality data is entered into the database in the first place, protecting the integrity of the data as much as possible. This is considered an entity relationship restriction because it restricts the types of data in any given entity.

Cardinality is about the row-to-row relationship between two table entities. There are three types of cardinality. The first is a one-to-one relationship, as shown in *Figure 14.2*:

Employee
EmployeeID
EmployeeName

Contact Info
EmployeeID
EmployeeAddress

Figure 14.2 – One-to-one cardinality

In *Figure 14.2*, each row in the first table corresponds to one row in the second table. This makes sense. Every employee has a unique ID and a legal address. In other words, every row in the `Employee` table corresponds to one row in the `Contact Info` table. This is a direct relationship. The next type is a one-to-many relationship, as shown in *Figure 14.3*:

Employee
EmployeeID
EmployeeName

Contact Info
EmployeeID
EmployeePhoneNumber

Figure 14.3 – One-to-many cardinality

Here, in *Figure 14.3*, each row in the first table corresponds to multiple rows in the second table. This looks very familiar, doesn't it? The only difference is that instead of `EmployeeAddress`, we have `EmployeePhoneNumber`. This is because, while each employee only has one mailing address, they may have multiple phone numbers: home phone, cell phone, office phone, and so on. This means for every employee in the first table, you might have three or four rows of `EmployeePhoneNumber` in the second table, all of which reference the same `EmployeeID`. Finally, the last type is a many-to-many relationship, as shown in *Figure 14.4*:

Employee
EmployeeID
EmployeeName

Sales
EmployeeID
SalesID

Figure 14.4 – Many-to-many cardinality

In *Figure 14.4*, we see a many-to-many relationship. An employee can make multiple sales, and each sale may have more than one employee working on it. That means that for every `EmployeeID`, you should have several `SalesIDs`, and every `SalesID` will have multiple `EmployeeIDs`. Many-to-many relationships can get messy and complicated rather quickly. The majority of restrictions around cardinality focus on how many relationships each row can have.

Summary

This chapter has covered a lot of information about data governance, including different levels and different aspects. First, we covered data security with different forms of access requirements, encryptions, transmissions, and data masking. Then, we talked about the Data Use Agreement contracts and their main sections: acceptable use, data processing with a focus on data breaches, data deletion, and data retention. Next, we went over different data classifications, such as PII, PHI, and PCI. Finally, we wrapped up our talk about data governance with a section on entity relationship requirements, which briefly touched on record link restrictions, data constraints, and cardinality.

In the next chapter, we will cover everything you need to know about data quality and management for the exam! I'll see you there!

Practice questions

Let's try to practice the material in this chapter with a few example questions.

Questions

1. Data4U, a **Software as a Service (SaaS)** company, would like to create a small dataset that includes de-identified data about how their clients have improved with the use of their software. They want every sales representative in the company to have access to this data to show to potential clients. What form of access requirement is this?

    A. User group-based

    B. Encryption-based

    C. Role-based

    D. Transmission-based

2. Which part of the Data Use Agreement includes explicit details about how the data is *not* supposed to be used?

    A. Acceptable use policy

    B. Data processing

    C. Data deletion

    D. Data retention

3. If you suspect that a data breach has occurred, which of the following is an appropriate response?

   A. Delete the data

   B. Ignore it

   C. Inform the impacted parties

   D. Do nothing

4. Which of the following variables would be considered PII?

   A. Geolocation

   B. Social media post

   C. Social Security number

   D. All of the above

5. A filter that only allows a specific kind of data to be entered into a dataset can be considered which kind of entity relationship restriction?

   A. Record link restriction

   B. Data constraint

   C. Cardinality

   D. None of the above

Answers

Now we will briefly go over the answers to the questions. If you got one wrong, make sure to review the topic in this chapter before continuing:

1. The answer is: Role-based

   The company does not care what group or department the person is in; they want every person in the company who has the role of sales representative to have access to this data. If they didn't care about the role and asked for everyone in the sales department to have access, it would have been user group-based.

2. The answer is: Acceptable use policy

   Remember that this section not only outlines ways the data can be used but also ways it should never be used.

3. The answer is: Inform the impacted parties

    Always report any suspected data breaches, and always make sure that the impacted parties are informed that their data may have been stolen.

4. The answer is: All of the above

    All of these pieces of information can be used to identify a specific individual, making them all PII.

5. The answer is: Data constraint

    Here, we are limiting the type of data that can be entered or stored in a specific data object, which makes this a data constraint.

15
Data Quality and Management

This chapter is all about making sure that you have high-quality data that is easy to access and use. Instead of focusing on policies, such as data governance, these are standards and techniques used to make sure that you have the best data possible because inaccurate data can only produce inaccurate results. First, we will go over what quality control is, when you should use it, and what sorts of things it is checking. Next, we will talk about specific methods for quality control. Finally, we will cover the concept of master data management and when to apply it.

In this chapter, we're going to cover the following main topics:

- Understanding quality control
- Validating quality
- Understanding master data management

Understanding quality control

Quality control is the process of testing data to ensure data integrity. Here, we will go over when to perform quality control checks and what sorts of things these checks are trying to find. Remember, bad data leads to bad results, and bad results are worse than no results because they are actively misleading. It is important that your data is as accurate as possible, and to do that, you need quality control.

When to check for quality

While you will probably automate as much of the quality control practices as you can, there are times beyond the routine when it is important to check the quality of your data. You may use different quality control techniques in different instances, but in general, you need to check your data any time there is a major change. Lots of things may qualify as a major change, but the most common are as follows:

- **Data acquisition**

  Data acquisition is whenever you get new data. This doesn't necessarily mean you are adding an observation to an old dataset, but every time you or your company get a new dataset. Before you use this new data, you have to check it. You must, whenever possible, ensure that the data has not been gathered or entered in such a way that would inherently introduce bias. It can be the cleanest, prettiest dataset in the world, but if it was collected in such a way that it is inherently biased, there is little you can do. At best, if you know how it is biased, you can attempt to account for it, and at worst, the data is useless. The other thing you need to do is check the current state of the data, which will be a more standard quality control check.

- **Data transformation**

  Data transformations are any time the data is changed from one form to another. There are lots of things that count as transformations, including intrahops, pass-throughs, and conversions – for example, if you attempt to normalize the data, or when you reformat a variable to match the same format as a variable from another table. It can also be something as small as changing the units you are using to report or changing the time zone. All of these transformations should be done in a new variable; try to avoid transforming the original data, or at least make sure the data is backed up. In this situation, you are mainly testing to see whether the transformation was performed correctly and that all of the data is still accurate.

- **Data manipulation**

  Data manipulation is when you change the shape of the data, not its content. You can think of this as data wrangling. It can include breaking one variable down into multiple variables, combining multiple variables into one, changing the level of data, or anything else that will make the spreadsheet look different without actually changing the values. Things like combining variables or breaking them down look like you are changing the data, but the same data is there, it is just represented differently. For example, you have a variable, `FullName`, but you want to break it down into two variables, `FirstName` and `LastName`, or vice versa. The name is exactly the same, it is just stored in one cell or two. Like with transformations, you are just making sure the manipulation went smoothly, so you are looking to see whether any new errors or mistakes were introduced.

- **Final product**

  The final product is one more check before your report. After you have run your analysis, this is just going back and making sure you didn't goof before your dashboard goes live to hundreds or thousands of people, or before you hand in the technical report you spent months on to the CEO.

  These are the major times to check the quality of your data outside of routine maintenance, but these are not the only times you can check your data. If you are in doubt, it never hurts to double-check. It is better to be safe than sorry.

Data quality dimensions

Okay, you now know when to check, but what exactly are you checking? The experts at CompTIA have identified a few key dimensions or specific things for which you should look. They are as follows:

- Data consistency

- Data accuracy

- Data completeness

- Data integrity

- Data attribute limitations

Data consistency is making sure your data is the same. This can apply to different levels. Within a variable, consistency is making sure it is reported the same way every time. Within a database or broader system, it means that if the same data is stored in multiple places, they should match. No matter the level, you are effectively making sure the data is uniform and things are the same throughout.

Data accuracy is whether or not the data is correct. This is, as you might guess, important. However, it is not always the easiest thing to check. A lot of this falls into the process of how it is collected, but whenever possible, it is best practice to check your data against an outside source, or go back to the original source to make sure the information you have stored is accurate, or whether there have been any changes.

Data completeness is checking for gaps. These gaps can come in multiple forms. Occasionally, it identifies an entire variable that should have been collected with the rest of the data or is required for a specific analysis. More often, this is looking for missing values within a variable that was collected. This goes right back to missing data, as discussed in *Chapter 4, Cleaning and Processing Data*.

Data integrity actually includes consistency, accuracy, and completeness, but it is slightly more than that. In certain fields, such as pharmaceuticals, data integrity has very defined rules and definitions, including making sure you know who entered every piece of data, when, and how. If you want to work with pharmaceutical data, make sure you look up these rules in detail. Luckily, for the exam, you only need to know that data integrity as a dimension is looking at consistency, accuracy, completeness, and security at a high level. This is less about specific data values and more about whether the process produces and maintains high-quality data.

Data attribute limitations is just another way to say data constraints, as we discussed in the previous chapter. Here, you are checking to make sure that only valid data can be entered in a given field.

Data quality rules and metrics

Data quality rules and metrics are guidelines that are created by the company, usually whoever is in charge of data governance. These are, effectively, cutoff scores and format templates that say clearly that this data is acceptable, and this data is not. A lot of these rules focus on conformity or making sure that all data of a certain type is in a specific format with a specific size. In other words, conformity, or non-conformity, makes up the rules that govern data consistency. Other rules, such as rows passed or rows failed, give clear numerical ranges for what is considered usable data and what isn't. If so many rows meet the guidelines and expectations, the data can be used; if not, it can't.

This gives you a general idea of what quality control is and when it should be applied. Next, we will look at a few methods for actually checking the quality of the data.

Validating quality

There are many ways to check the quality of your data, but there are a few forms that are more common than others. Let's take a look at the most popular, which are also the ones you need to know for the exam. These include the following:

- Cross-validation
- Sample/spot check
- Reasonable expectations
- Data profiling
- Data audits

Some of these are pretty self-explanatory, but let's go into a little more detail for each.

Cross-validation

Cross-validation is a statistical analysis that checks to see whether the results of a different analysis can be generalized. This analysis has many different uses. It can check data model effectiveness, specifically if you are checking for overfitting. Often, it is used to figure out what the hyperparameters should be for your model, and is a great tool for reducing test error. Cross-validation is a useful tool that can be applied in several different ways to check the quality and function of your data and data models.

Sample/spot check

Sample/spot checks are usually quick checks that focus on one or two data quality dimensions. They can be scheduled and held at regular intervals or after you change the data in any meaningful way, but often, these happen when you notice something unusual with the data while you are working with it. Maybe you get an error or a result that doesn't make any sense and you go back to look at the raw data to see whether there is something wrong with the data itself.

Reasonable expectations

Reasonable expectations are simply whether or not the data makes any sense. This can come in a few different forms. Sometimes, this is just a gut check to see whether or not the data is anywhere near what you expect it to be. Occasionally, this is a more formalized process, where professionals decide what are considered "reasonable" values, and an error or flag goes off every time the data goes outside of those boundaries. If you work for a company that races snails professionally, and all of the snails in the past have weighed between 3.66 g and 7.86 g, you should be concerned if suddenly, all of the new values are over 1,000 g. Some major change has taken place, probably an error, and not that all snails have grown over 1 kg, but you need to investigate.

Data profiling

Data profiling is a formal process applied to entire databases that checks for data quality and problems. While we call this a process, it is usually divided into multiple processes that each check specific aspects of the database. There are several types, but the main three are as follows:

- Structure discovery
- Content discovery
- Relationship discovery

These each focus on their namesake: structure, content, and relationship, respectively. Data profiling can check how efficiently your database is laid out, what types of data it contains, and even how the different data objects are connected. This kind of high-level database function quality checks only comes from data profiling. Data profiling is important but is rarely performed by a data analyst, usually being left in the hands of database administrators or other database specialists.

Data audits

Data audits check to see whether a dataset is appropriate for a specific goal. They are systematic and usually performed on a schedule. For example, products that release information on a weekly or monthly schedule will almost always have a data audit once for every release cycle. They should also be performed at every stage of the data life cycle. This process takes a lot of time and is often left to a quality assurance analyst or something similar.

Automated checks

Of course, it is best to automate the quality checks as much as possible. You will always need human eyes on it at some point, but the more you can automate, the fewer hours you have to spend checking and double-checking countless rows and columns. There are many different things you can automate to make your life easier, but the exam wants to point out two specifically:

- Data field to data type validation
- Number of data points

Data field to data type validation is simply an automated check to ensure that values in a field are the expected data type, size, or within the expected value range. Alternatively, this can be a data constraint, which stops inappropriate data from ever being entered into the field.

The number of data points is a way to automate completeness checks. If you have 1,000 observations in a dataset, but a specific variable only has 500 data points, that means that half of the values for that variable are missing. It is a quick check and easy to automate.

There are many approaches to performing quality control checks, but there is no one perfect solution. The only way to ensure that you have access to quality data is to use a combination of these methods based on your particular circumstances. Now that you have an idea about how to approach data quality, let's move on to master data management.

Understanding master data management

Master data management (MDM) is effectively the process of creating and managing a centralized data system. The idea is that you create a golden record, or a single source of truth. Having one place where all of your data is clean, standardized, consolidated, and up to date has many benefits. You end up with higher data quality, higher data integrity, cleaner data, and much faster and easier access to the data, and you can even format the data so that it is ready to report. Also, if you create new data objects, you can connect them to the golden record and automatically populate the fields with the appropriate data. You can even automate compliance checks. Effectively, you have the data in its highest quality on tap and ready to use by anyone who has access.

That said, MDM is not a practice that is used by every company. It sounds great, but it can also be a lot of work and expensive to set up. There are even variations, where MDM is only practiced on specific types of data. Maybe a company only has a golden record for their customer data or PII, and everything else is left separate. Alternatively, they can have multiple golden records, each for a different department or data type. Setting up and maintaining MDM is not the easiest thing to do, but there are specific tools and software to help.

When to use MDM

While many use MDM for its many benefits, there are specific times when it comes in handy. These times are as follows:

- **Mergers and acquisitions**

  Mergers and acquisitions happen when two companies combine, or one company buys another company out. Times like this are always a struggle when it comes to data. The two companies probably had very different data that was formatted differently and stored in a database with a very different structure. MDM helps by creating a single, consistent record from these two sources that is easy to access and use.

- **Policy**

  Policy is in reference to compliance. Having all of the records nice, neat, and in one place makes it much easier to check to see whether you meet all regulations and restrictions. This is especially important when you deal heavily with protected data such as PII, PHI, or PCI. If you work with a lot of protected data, or data with several regulations, MDM is a great way to protect yourself.

- **Streamlining data access**

  Streamlining data access means that you can get the data you need faster. All of the data you need can be accessed from a single table. No joins or fancy queries are required with MDM. It just makes it easier.

Processes of MDM

Okay, you have a rough idea of what MDM is and when it is useful, but let's talk briefly about what the process looks like. While it can be a long and complicated process, we've broken it down into a few simple, generalized steps:

- **Consolidation**

  Consolidation is the actual creation of the golden record. Here, you are combining data from multiple sources into one place. Note that the separate original tables still exist, and any updates to the golden record automatically update data sources. This way, if you have the same variable in multiple places, you don't need to update them separately or worry that they might not be consistent. Updating the golden record will update every data source so that they all say the exact same thing.

- **Standardization**

  Standardization is making things uniform. This can be standardizing field names between tables, standardizing units, standardizing formats, or even standardizing data entry regulations. All of your data is on the same page literally, so it is time to get it on the same page figuratively. This makes sure that all of your data works together and can be used as a whole.

- **Data dictionary**

  Data dictionaries are documents that give definitions and attributes for every variable in the dataset. They can also include information on structure, relationships, and data organization. Data dictionaries are one of the most important tools a data analyst can have, and they should be used whether or not you practice MDM. Databases are rarely the property of one person. Often, you have multiple people using a database, or at least, the database will be handed off to someone else when you leave. Without proper documentation, no one will understand what the data is or how it should be used. This is especially important if your data uses shorthand for variable names. For example, RNSVBQ17 means absolutely nothing by itself, but a dictionary can tell you that this variable holds patient responses to the Registered Nurse Survey Version B Question 17. Hopefully, it will even tell you what that question was and how it relates to other variables. Data dictionaries can be handwritten notes or software-generated documents, it doesn't matter; just make sure that one exists.

Summary

This chapter has gone over some important information for ensuring the highest possible quality of your data. First, we covered what quality control is, when it should be applied, and what sorts of things it looks into. Then, we looked at some common methods and approaches to quality control. Finally, we went over what MDM is, when it should be applied, and the processes behind it.

Congratulations! This wraps up the educational content of this book. You have now covered every concept that will be on the exam. There is a lot of information covered in this book, so feel free to go back and review any sections with which you had any problems. Next, it's time for the practice exams!

Practice questions

Let's try to practice the material in this chapter with a few example questions.

Questions

1. Which of the following is an appropriate time to check the quality of your data?

    A. After data manipulation

    B. After data transformation

    C. Before the final report

    D. All of the above

2. Which of the following is a data quality dimension?

    A. Data completeness

    B. Data retention

    C. Rows passed

    D. Data manipulation

3. Which of the following is a structured formal process for identifying the quality and efficiency of an entire database?

    A. Cross-validation

    B. Spot check

    C. Reasonable expectations

    D. Data profiling

4. **Over 9,000!** is a power company that is rapidly growing in its area. Recently, it purchased another power company that was one of its major competitors. This is an appropriate time to institute MDM. True or false?

    A. True

    B. False

5. Creating a document that explains what the variables in a dataset are, how they are used, and how they connect to one another represents which part of the MDM process?

    A. Data audits

    B. Consolidation

    C. Data dictionary

    D. Standardization

Answers

Now we will briefly go over the answers to the questions. If you got one wrong, make sure to review the topic in this chapter before continuing:

1. The answer is: All of the above

    These are all circumstances where you should check for quality.

2. The answer is: Data completeness

    Data completeness is the only data quality dimension listed. Data retention is a section in the data use agreement, rows passed is an aspect of a data quality rule, and data manipulation is a general concept of data shaping.

3. The answer is: Data profiling

    While these are all methods of validating quality, data profiling is the only structured approach to checking the quality of an entire database.

4. The answer is: True

    One company buying another is called an acquisition, and acquisitions and mergers are one of the major reasons for MDM. This is because you have two completely separate datasets, one from each company, and you have to merge them into one dataset that is still somehow usable.

5. The answer is: Data dictionary

    A data dictionary is, as you would expect, a dictionary of your data. It gives definitions for every variable, as well as how they are used and how they relate to other variables. These are very useful tools and an important part of the MDM process.

Part 4: Mock Exams

Now, it's time to put the learning into practice through a practice exam, complete with full explanations of the answers with context and tools to improve learning.

This part covers the following chapters:

- *Chapter 16, Practice Exam One*
- *Chapter 17, Practice Exam Two*

16
Practice Exam One

This practice exam will give you the correct number of questions in the correct proportions. There are certain types of questions or interactive dashboards that cannot be simulated in a book, but this exam will give you a good idea of what material you know well and what material you may need to review. If you know the material well, the format of the questions should not be a problem. Also note that CompTIA is constantly updating and adding new questions on the same topics, written by different subject matter experts. Some of these questions will be confusing, poorly written, and may not even follow the CompTIA guidelines. They have to test questions to see whether they work, and they don't always. Expect a weird question or two, and don't let it throw you off.

To make this practice exam as useful as possible, you should follow the testing experience as closely as possible. Make sure you are in a quiet and clean environment before you commence and have access to a simple calculator, a timer, and the ability to take notes.

You have 90 minutes to answer 90 questions, so you will need to average 1 question per minute. That said, some questions are quick and easy, while others will take several minutes. If you are stuck on a question, it may be best to guess and make a note to come back to it if you have time. All unanswered questions are counted the same as an incorrect answer, so it is important that you at least attempt to answer every question.

Take a deep breath.

You can do this.

Practice exam one

1. If you would like your dashboard to be delivered to the recipients only once on a specific date, what would be the most appropriate approach?

   A. Access permissions

   B. Subscription

   C. Interactive saved searches

   D. Scheduled delivery

2. A schema with denormalized tables would be what type of schema?

   A. Fact constellation schema

   B. Snowflake schema

   C. Star schema

   D. Galaxy schema

3. If you were to perform a right join on the following tables, what would be the result?

Left Table		Right Table	
ClientID	Name	ClientID	Name
1	Smith, Laurence	1	Austin, TX
2	Brown, Betty	2	Denver, CO
3	Hook, Phil	3	Tulsa, OK
4	Roark, Jen	7	Phoenix, AZ
5	Cox, Jona	8	Seattle, WA
6	Humbert, Ren	9	Baltimore, MD

A.

Joined Table		
ClientID	Name	City
1	Smith, Laurence	Austin, TX
2	Brown, Betty	Denver, CO
3	Hook, Phil	Tulsa, OK
4	Roark, Jen	NULL
5	Cox, Jona	NULL

Joined Table		
ClientID	Name	City
6	Humbert, Ren	NULL
7	NULL	Phoenix, AZ
8	NULL	Seattle, WA
9	NULL	Baltimore, MD

B.

Joined Table		
ClientID	Name	City
1	Smith, Laurence	Austin, TX
2	Brown, Betty	Denver, CO
3	Hook, Phil	Tulsa, OK

C.

Joined Table		
ClientID	Name	City
1	Smith, Laurence	Austin, TX
2	Brown, Betty	Denver, CO
3	Hook, Phil	Tulsa, OK
4	Roark, Jen	NULL
5	Cox, Jona	NULL
6	Humbert, Ren	NULL

D.

Joined Table		
ClientID	Name	City
1	Smith, Laurence	Austin, TX
2	Brown, Betty	Denver, CO
3	Hook, Phil	Tulsa, OK
7	NULL	Phoenix, AZ
8	NULL	Seattle, WA
9	NULL	Baltimore, MD

4. Which of the following represents the count of observations that falls into each category?

    A. The confidence interval

    B. The z-score

    C. The percent difference

    D. Frequency

5. Forecasting falls into what general type of analysis?

    A. Performance analysis

    B. Link analysis

    C. Exploratory data analysis

    D. Trend analysis

6. The following is a representation of what type of visualization?

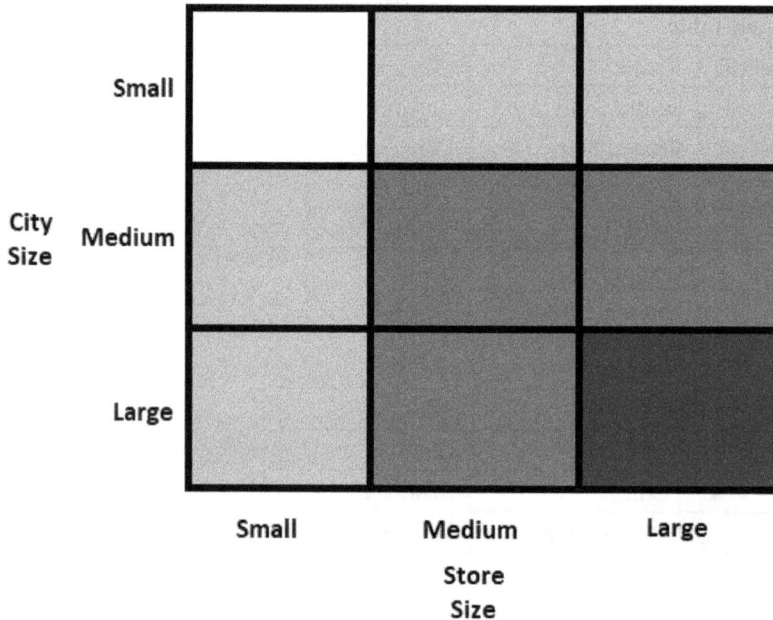

    A. A tree map

    B. A heat map

    C. A geographic map

    D. A stacked bar chart

7. Which of the following analytical tools is considered a programming language?

   A. Dataroma

   B. Python

   C. Minitab

   D. Qlik

8. A person's credit card information is considered what type of protected data?

   A. PII

   B. PHI

   C. PCI

   D. PBI

9. Your manager would like to know whether an employee's height has any relationship to how productive they are before the end of the week, so they can decide whether they should hire taller employees. What type of report is most appropriate in this scenario?

   A. A self-service report

   B. A recurring report

   C. A research report

   D. An ad hoc report

10. Which of the following would contain the true mean?

    A. The confidence interval

    B. The interquartile range

    C. The range

    D. The standard deviation

11. When you are creating a new dashboard, which of the following should be done first?

    A. Deliver the dashboard

    B. Get approval

    C. Create the dashboard

    D. Create a mockup/wireframe

12. You are asked to create a visualization for a sales report. This single visualization looks at revenue from each division in the company. How much revenue came from each department within each division is also a requirement. What visualization would be most appropriate?

   A.   A bubble chart

   B.   A pie chart

   C.   A stacked bar chart

   D.   A line chart

13. The following table is in chronological order, with new values added to the bottom. This table is an example of updating a table by what means?

Total Number of Beans	Red Beans	Blue Beans	Yellow Beans
10	X	X	X
12	X	X	X
11	3	5	3
9	2	6	1
10	4	3	3

   A.   Adding variables

   B.   Removing variables

   C.   Deleting historical data

   D.   Active record

14. The following dataset is an example of what type of error?

City
los angeles
LA
Los Angeles
Los Angelus
la
los angeles

   A.   Data type validation

   B.   Specification mismatch

   C.   Redundant data

   D.   Invalid data

15. If you are trying to explain the complicated relationship between different factors to an audience that is not technical, such as the general public, what type of visualization would be most appropriate?

    A. Heat map

    B. Histogram

    C. Infographic

    D. Bubble chart

16. What does ETL stand for?

    A. Exact, time, line

    B. Extract, transform, load

    C. Electronic, transfer, logistics

    D. Extra, transactional, Lambda

17. The following screenshot represents what type of survey question?

    5. Chunky peanut butter is, objectively, better. ⚲ 0

    ◯ Strongly agree

    ◯ Agree

    ◯ Neither agree nor disagree

    ◯ Disagree

    ◯ Strongly disagree

    A. Likert

    B. Drop-down

    C. Multiple choice

    D. Text-based

18. Which section of a data use agreement includes information on what happens if consent is withdrawn?

    A. The acceptable use policy

    B. Data processing

    C. Data deletion

    D. Data retention

19. If your goal was to have a database that required the fewest number of joins possible, which schema would be most appropriate?

    A. A snowflake schema

    B. A star schema

    C. A snowball schema

    D. A galaxy schema

20. The following dataset is an example of what type of error?

Employee ID	LastName	FirstName	Department	YearsWithCompany
83784	Benhill	Floyd	Sales	12
64986	Chane	Jill	IT	1
64986	Chane	Jill	IT	1
64986	Chane	Jill	IT	1
93671	Hanson	Richard	HR	15
37816	Smith	Trudy	Sales	21

    A. Duplicate data

    B. Redundant data

    C. Missing data

    D. Invalid data

21. Which of the following is something to consider when checking for data quality?

    A. Data visualization

    B. Data transmission

    C. Data accuracy

    D. Data encryption

22. Which data validation approach should you take if you need to see whether the results of an analysis can be generalized?

    A. Data auditing

    B. Data profiling

    C. Spot checking

    D. Cross-validation

23. What type of analysis gives you basic information about the shape and structure of your data before you even start using it?

    A. Performance analysis

    B. Trend analysis

    C. Exploratory data analysis

    D. Link analysis

24. A small tech start-up is trying to decide between two designs for the same product: Design A and Design B. They have run several trials on both designs, and they think that Design A is more efficient, but they aren't sure. In this scenario, what is the alternative hypothesis?

    A. There is no significant difference between Design A and Design B

    B. There is a significant difference between Design A and Design B

    C. There is a difference between Design A and Design B, but it is not significant

    D. It doesn't matter which design they pick

25. A project manager requests information on KPIs to see whether their team is staying on schedule. What type of analysis is this?

    A. Trend analysis

    B. Performance analysis

    C. Exploratory data analysis

    D. Link analysis

26. Find the variance of the following dataset:

$$56, 46, 27, 31, 40$$

    A. 11.6

    B. 40.0

    C. 112.8

    D. 135.5

27. Using the independent variable to predict the dependent variable best describes what analysis?

    A. A t-test

    B. A simple linear regression

    C. A chi-square

    D. A z-score

28. The following represents what type of distribution?

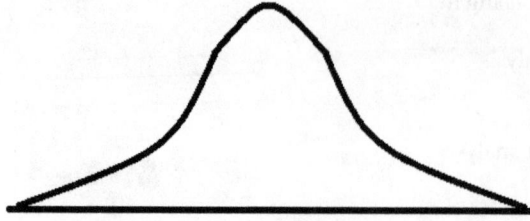

   A. Normal
   B. Uniform
   C. Exponential
   D. Bernoulli

29. Which of the following has a direct impact on how fast and efficient a dashboard is?

   A. Access permissions
   B. Interactive saved searches
   C. Subscription
   D. Scheduled delivery

30. Which of the following file types is used to pass data through websites without having anything to do with the structure of a website?

   A. JSON
   B. HTML
   C. JS
   D. XML

31. Find the mean of the following dataset:

39, 37, 42, 39, 43

   A. 37
   B. 39
   C. 40
   D. 42

32. Which of the following data storage solutions is most appropriate for large amounts of unstructured data?

    A. Data mart

    B. Data lake

    C. Data warehouse

    D. Data puddle

33. If you wanted to compare the batting records of two baseball athletes over the course of a year, which of the following analyses would be most appropriate?

    A. Correlation

    B. Z-score

    C. T-test

    D. Chi-square

34. You are working with the following dataset of data gathered from different robots attempting an obstacle course. The team wants to know whether the weight of the robot had any impact on how far it got in the course. Which variables would be considered appropriate data content for the report?

Date	Time(s)	Vol	Ins	Res	Operator	Distance(m)	RobotWeight(lb)
7/5/2019	33	1.9470	471	40	Oscar Mayor	22	9
7/12/2019	35	1.3958	581	86	Oscar Mayor	25	11
7/19/2019	44	1.3905	304	4	Oscar Mayor	23	10
7/26/2019	30	1.5938	482	15	Oscar Mayor	24	8
8/2/2019	43	1.1938	561	51	Oscar Mayor	19	11
8/9/2019	33	1.8404	309	27	Oscar Mayor	21	9
8/16/2019	35	1.4893	701	83	Oscar Mayor	16	9
8/23/2019	40	1.2843	824	79	Oscar Mayor	15	10
8/30/2019	35	1.9837	754	73	Oscar Mayor	22	11
9/6/2019	43	1.7489	601	48	Oscar Mayor	24	9

    A. Time(s), Operator, and Distance(m)

    B. Date, Vol, Res, Distance(m), and RobotWeight(lb)

    C. Date, Distance(m), and RobotWeight(lb)

    D. Every variable

35. An analysis that compares a sample to the population to see whether it is a good representation best describes what analysis?

    A. Correlation

    B. T-test

    C.   Simple linear regression

    D.   Chi-square

36. Which data quality dimension means ensuring your data is all reported in the same format?

    A.   Accuracy

    B.   Completeness

    C.   Consistency

    D.   Attribute limitation

37. Customer-facing agents in your company have requested a report that gives them the most up-to-date customer information and rates possible. What type of report is most appropriate in this scenario?

    A.   A point-in-time report

    B.   A static report

    C.   A dynamic report

    D.   A research report

38. Which of the following would you find in a structured database?

    A.   Images

    B.   Defined rows/columns

    C.   Video files

    D.   Social media data

39. If you are updating an old report by rerunning the analyses on updated data, which of the following values will change?

    A.   The reference data sources

    B.   The appendix

    C.   The reference dates

    D.   The FAQs

40. When choosing fonts for your report, what should be your first consideration?

    A.   Font size

    B.   The number of fonts

    C.   Font types

    D.   Branding

41. The sales department of your company would like to be able to look up the latest sales data at any point in time so they can make decisions quickly about setting rates for new contracts. What type of report is most appropriate in this scenario?

    A. A research report

    B. An ad hoc report

    C. A self-service report

    D. A recurring report

42. Deleting an entire row of data because of one value in it is what type of deletion?

    A. Pairwise deletion

    B. Listwise deletion

    C. Variable deletion

    D. Filtering

43. Code or software that tells you information about your environment or file paths is an example of what?

    A. System functions

    B. Conditional operators

    C. Recoding

    D. Transposition

44. if, and, or, and not are examples of what?

    A. Dummy coding

    B. Transposition

    C. Reduction

    D. Conditional operators

45. What do you call the process of filling gaps in the data with the average of the values that are present?

    A. Interpretation

    B. Infusion

    C. Imputation

    D. Interpolation

46. The following depicts what kind of visualization?

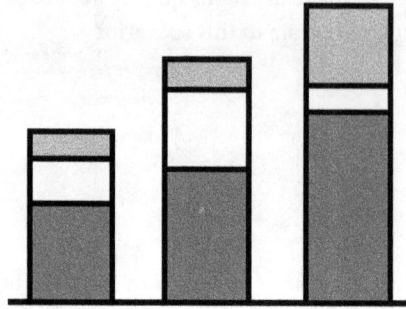

   A. A waterfall chart

   B. A histogram

   C. A bar chart

   D. A stacked bar chart

47. Which of the following are circumstances under which you should check the quality of your data?

   A. Data transformation

   B. Data transmission

   C. Data encryption

   D. Data deletion

48. The following graph depicts what kind of correlation?

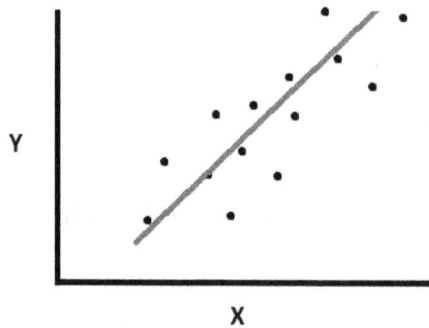

   A. No correlation

   B. Negative correlation

   C. Positive correlation

   D. Semi-correlation

49. You receive your results for an exam, and the results of the other students are posted anonymously. You want to know how your exam results compare to the normal distribution of your classmates' scores. Which analysis is most appropriate?

    A. Chi-square

    B. T-test

    C. Z-score

    D. Simple linear regression

50. The following dataset is an example of what type of data?

Total Number of Beans	Red Beans	Blue Beans	Yellow Beans
10	3	4	3
12	2	6	4
11	3	5	3

    A. Unstructured

    B. Relational

    C. Non-relational

    D. Semi-structured

51. A manager with little to no experience working with data would like access to the data so that they can test some things out for themselves. What is the most appropriate business requirement to consider when making this report?

    A. Views

    B. Data range

    C. Filters

    D. Frequency

52. You calculate that the average number of clicks per minute for the year for a small e-commerce website is 12.8. Last year, you calculated this to be 8.7. How much has the average number of clicks per minute increased since last year?

    A. 35%

    B. 38%

    C. 47%

    D. 52%

53. Compliance reports are generally considered what type of report?

    A. An ad hoc report

    B. A research report

    C. A recurring report

    D. A self-service report

54. What is redundant data?

    A. The same information recorded in multiple columns

    B. Data that is incomplete or blank

    C. The same information recorded in multiple rows

    D. Data that does not meet formatting requirements

55. The following visualization represents what type of distribution?

    A. Normal

    B. Uniform

    C. Non-parametric

    D. Exponential

56. An IP address is considered what type of protected data?

    A. PII

    B. PHI

    C. PCI

    D. PIFI

57. You run a simple AB study to compare two different advertising campaigns. Assuming an alpha of 0.05, which of the following p-values would cause you to reject the null hypothesis?

    A. 0.3

    B. 0.5

    C. 0.06

    D. 0.008

58. If you had a variable called `BirdCount` that kept track of the number of birds that flew by your window on a specific day, what variable type would it be?

    A. Nominal

    B. Continuous

    C. Ordinal

    D. Discrete

59. The act of automatically collecting, processing, and storing online transactions is called what?

    A. OLAP

    B. ETL

    C. ELT

    D. OLTP

60. Breaking large chunks of data down into small, usable pieces is called what?

    A. Interpretation

    B. Parsing

    C. Reduction

    D. Interpolation

61. A new variable that specifically holds a calculation of other variables is called what?

    A. A derived variable

    B. A binary variable

    C. A variable deletion

    D. An ordinal variable

62. The marketing team for your company has created a sample of the company's customers, but before they run any tests, they want to know whether the sample accurately reflects the larger population of customers. Which of the following analyses is most appropriate?

    A. The chi-square test for independence

    B. The chi-square test for homogeneity

    C. The chi-square goodness of fit

    D. The chi-square test for linearity

63. A sales manager believes that there is a connection between the customer's age and how likely they are to purchase the product. What type of analysis would be most appropriate?

    A. Trend analysis

    B. Performance analysis

    C. Link analysis

    D. Exploratory data analysis

64. You are asked to create a visualization that tracks ROI over the past 4 years, looking for general trends. What is the most appropriate visualization?

    A. A line chart

    B. A scatter plot

    C. A heat map

    D. A bubble chart

65. Where on a dashboard do you put instructions for how to use it?

    A. The appendix

    B. The cover page

    C. The FAQs

    D. The reference dates

66. What conclusion can you draw from the following visualization?

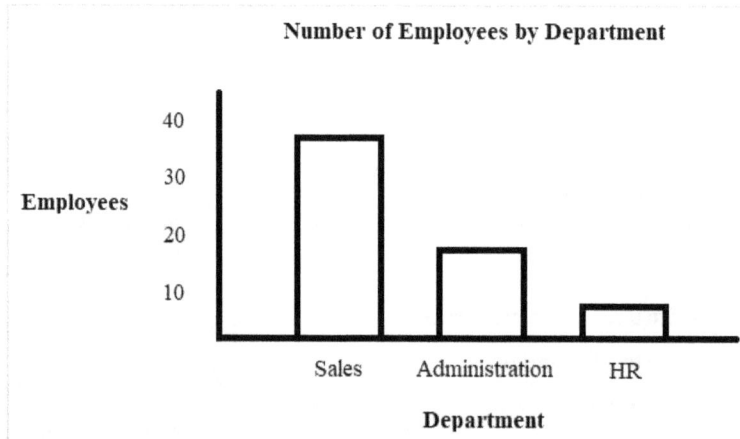

Number of Employees by Department

    A. This accounts for every employee in the company

    B. Administration is the smallest department

C. Administration is the biggest department

D. There are at least 60 employees in the company

67. The following graph represents what kind of visualization?

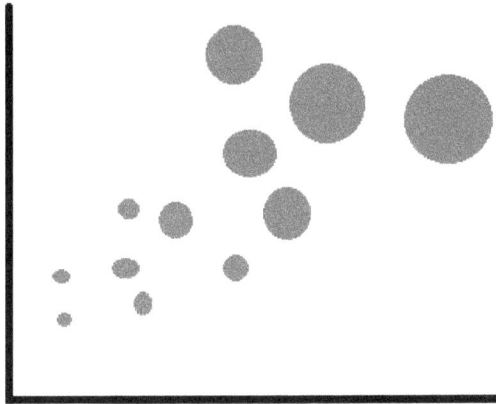

A. A waterfall chart

B. A scatter plot

C. A pie chart

D. A bubble chart

68. A data point that is so much smaller than every other data point in the dataset that it drastically lowers the average for the entire dataset is an example of what?

A. A specification mismatch

B. Duplicate data

C. An outlier

D. Data type validation

69. Which section of a data use agreement includes information on the consequences of using the data improperly?

A. The acceptable use policy

B. Data processing

C. Data deletion

D. Data retention

70. The following is an example of what type of join?

Left Table			Joined Table				Right Table	
ClientID	Name		ClientID	Name	City		ClientID	City
1	Smith, Laurence		1	Smith, Laurence	Austin, TX		1	Austin, TX
2	Brown, Betty		2	Brown, Betty	Denver, CO		2	Denver, CO
3	Hook, Phil		3	Hook, Phil	Tulsa, OK		3	Tulsa, OK
4	Roark, Jen						7	Phoenix, AZ
5	Cox, Jona						8	Seattle, WA
6	Humbert, Ren						9	Baltimore, MD

A. Outer join

B. Inner Join

C. Left Join

D. Right Join

71. Which of the following instances is an ideal time to implement MDM?

A. When data is manipulated

B. When data is transferred

C. When a company is purchased

D. When data is transformed

72. In the following table, which variable is specifically there to indicate that a row is the most recent value?

Magic Number	Active Record	Active Start	Active End
41	No	11/11/2011	12/12/2012
42	Yes	12/12/2012	

A. Magic Number

B. Active Record

C. Active Start

D. Active End

73. A large data storage solution for relational data, focusing on efficiency and following a snowflake schema, would most likely be what?

A. A data warehouse

B. A data mart

C. A data lake

D. A data puddle

74. The following dataset is an example of what concept?

Month	UnitsSold	Color	Red	Blue	Yellow
August	432	Red	1	0	0
August	365	Blue	0	1	0
August	154	Yellow	0	0	1
September	398	Red	1	0	0
September	386	Blue	0	1	0
September	108	Yellow	0	0	1

A. Recoding a number into a category

B. Recoding a category into a number

C. Dummy coding

D. Transposition

75. What is the range of the following dataset:

$$25, 38, 50, 49, 38$$

A. 25

B. 38

C. 40

D. 50

76. The following diagram represents what type of database schema?

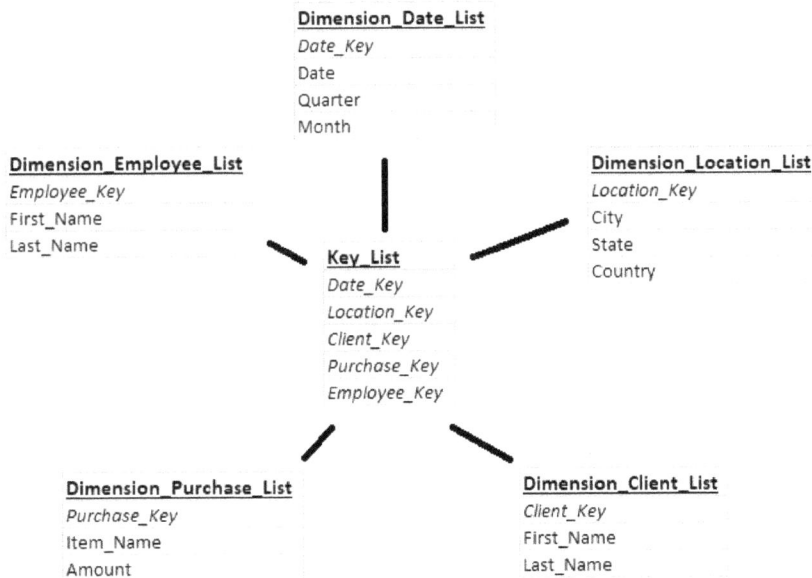

Dimension_Date_List
Date_Key
Date
Quarter
Month

Dimension_Employee_List
Employee_Key
First_Name
Last_Name

Dimension_Location_List
Location_Key
City
State
Country

Key_List
Date_Key
Location_Key
Client_Key
Purchase_Key
Employee_Key

Dimension_Purchase_List
Purchase_Key
Item_Name
Amount

Dimension_Client_List
Client_Key
First_Name
Last_Name

    A. A snowflake schema

    B. A star schema

    C. A galaxy schema

    D. A fast constellation schema

77. Your company has set up a database that can only be accessed by people with the job title data analyst. What type of access requirement is this?

    A. Data encryption

    B. User group-based

    C. Role-based

    D. Data transmission

78. A pharmaceutical lab wants to know whether a memory-enhancing drug works. They have two mice run the same maze multiple times, one with the medication and one without. You analyze the results and receive a p-value of 0.4. Assuming an alpha of 0.05, how do you interpret the results?

    A. Accept the alternative hypothesis and reject the null hypothesis

    B. Reject the alternative hypothesis and accept the null hypothesis

    C. Accept the alternative and null hypotheses

    D. Reject the alternative and null hypotheses

79. Which of the following is a prewritten query that allows the user to only enter very specific information to target data?

    A. Index

    B. Parameterization

    C. Filter

    D. Sort

80. What do you call the process of manually or automatically checking the data type of a variable to avoid errors?

    A. Data type validation

    B. Imputation

    C. Specification mismatch

    D. Interpolation

81. A piece of code that requests information from an API and then waits for a response before continuing has what sort of connection?

    A. Structured

    B. Unstructured

    C. Synchronous

    D. Asynchronous

82. The following tables have what sort of cardinality?

Employee Information		Sales Information	
EmployeeID	EmployeeName	EmployeeID	SalesID
8279	Steve	8279	826705
7904	Joan	8279	379001
2905	Frank	2905	758982
8391	Bertha	8391	178904
		7904	389057
		8279	890471
		8391	689409
		8391	789039
		8391	689315
		8279	492084
		2905	291046
		2905	164829

    A. One-to-one

    B. One-to-many

    C. Many-to-many

    D. There is no entity relationship

83. Which data validation approach should you take to see whether a dataset is appropriate for a specific goal?

    A. Reasonable expectations

    B. Data audits

    C. Data profiling

    D. Cross-validation

84. What conclusion can you draw from the following visualization?

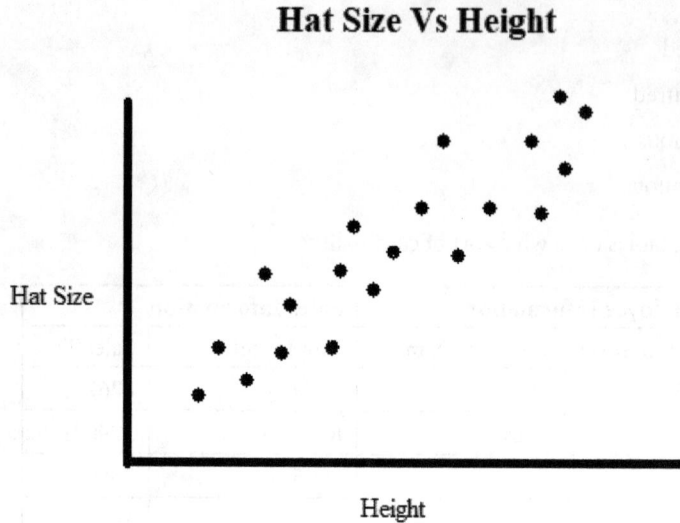

Hat Size Vs Height

Hat Size

Height

A. Being taller makes your head bigger

B. Having a small head makes you taller

C. The taller you are, the more likely you are to have a larger hat size

D. The taller you are, the more likely you are to have a smaller hat size

85. A WAV file holds what type of information?

A. Image

B. Video

C. Audio

D. Text

86. The process of using code to automatically collect data from websites for you is called what?

A. Web scraping

B. Web services

C. Non-relational

D. Semi-structured

87. The following dataset is an example of what type of data?

Total Number of Beans	Red Beans	Blue Beans	Yellow Beans
10	3	4	3
12	2	6	4
11	3	5	3

   A. Structured

   B. Relational

   C. Both structured and relational

   D. Neither structured nor relational

88. Which of the following is a key process of MDM?

   A. Data dictionary

   B. Data encryption

   C. Data transformation

   D. Data manipulation

89. The following depicts an action that has been performed on a dataset. What is the name of that action?

Index	Employee ID	LastName	FirstName	Department	YearsWithCompany
1	83784	Benhill	Floyd	Sales	12
2	64986	Chane	Jill	IT	1
3	93671	Hanson	Richard	HR	15
4	37816	Smith	Trudy	Sales	21

Index	Employee ID	LastName	FirstName	Department	YearsWithCompany
4	37816	Smith	Trudy	Sales	21
3	93671	Hanson	Richard	HR	15
1	83784	Benhill	Floyd	Sales	12
2	64986	Chane	Jill	IT	1

A. Filtering

B. Subsets

C. Parameterization

D. Sorting

90. A software company recently released an update to make their program faster, and they want to know whether it worked. You compare several trials using the software before and after the patch. Finally, you decide that the update did make the software faster. Unfortunately, you are wrong. What type of error is this?

A. Type I

B. Type II

C. Type III

D. Type IV

Congratulations!

You made it through your first practice exam! Take a moment to relax and celebrate. Get a cookie; you deserve it. Make sure to hydrate and take care of anything you couldn't do during the exam. Once you are set, compare your answers to the exam answers listed in the following section.

CompTIA says that a pass score is 675 on a scale of 100 to 900, but they do not release any information on grading or how they weight the different questions. Also, if they are testing a new question and everyone gets it wrong, it probably won't be counted against you. Overall, this makes grading a practice exam a pain. As a rough guideline, try to miss 20 or fewer questions. That said, what is important is that you note what topics you are struggling with and make sure you review them. It's okay if you miss more than 20 on your first try, but make sure you review those topics and understand them before you try again.

Practice exam one answers

1. The answer is: Scheduled delivery

    If you are unsure why, please review *Chapter 12, Reporting Process – Understanding Report Delivery*

2. The answer is: Star schema

    If you are unsure why, please review *Chapter 2, Data Structures, Types, and Formats – Going through the Data Schema and Its Types*

3. The answer is:

Joined Table		
ClientID	Name	City
1	Smith, Laurence	Austin, TX
2	Brown, Betty	Denver, CO
3	Hook, Phil	Tulsa, OK
7	NULL	Phoenix, AZ
8	NULL	Seattle, WA
9	NULL	Baltimore, MD

    If you are unsure why, please review *Chapter 5, Data Wrangling and Manipulation – Merging Data*

4. The answer is: Frequency

    If you are unsure why, please review *Chapter 8, Common Techniques in Descriptive Statistics – Understanding Frequencies and Percentages*

5. The answer is: Trend analysis

    If you are unsure why, please review *Chapter 6, Types of Analytics – Discovering Trends*

6. The answer is: A heat map

    If you are unsure why, please review *Chapter 13, Common Visualizations – Understanding Heat Maps, Tree Maps, and Geographic Maps*

7. The answer is: Python

    If you are unsure why, please review *Chapter 11, Types of Reports – Knowing Important Analytical Tools*

8. The answer is: PCI

    If you are unsure why, please review *Chapter 14, Data Governance – Understanding Data Classifications*

9. The answer is: An ad hoc report

    If you are unsure why, please review *Chapter 11, Types of Reports – Understanding Ad hoc and Research Reports*

10. The answer is: The confidence interval

    If you are unsure why, please review *Chapter 8, Common Techniques in Descriptive Statistics – Discovering Confidence Intervals*

11. The answer is: Create a mockup/wireframe

    If you are unsure why, please review *Chapter 12, Reporting Process – Understanding the Report Development Process*

12. The answer is: A stacked bar chart

    If you are unsure why, please review *Chapter 13, Common Visualizations – Comprehending Charts with Bars*

13. The answer is: Adding variables

    If you are unsure why, please review *Chapter 2, Data Structures, Types, and Formats – Updating Stored Data*

14. The answer is: Invalid data

    If you are unsure why, please review *Chapter 4, Cleaning and Processing Data – Understanding Invalid Data, Specification Mismatch, and Data Type Validation*

15. The answer is: Infographic

    If you are unsure why, please review *Chapter 13, Common Visualizations – Understanding Infographics and Word Clouds*

16. The answer is: Extract, transform, load

    If you are unsure why, please review *Chapter 3, Collecting Data – Differentiating ETL and ELT*

17. The answer is: Likert

    If you are unsure why, please review *Chapter 3, Collecting Data – Collecting Your Own Data*

18. The answer is: Data deletion

    If you are unsure why, please review *Chapter 14, Data Governance – Knowing Use Requirements*

19. The answer is: A star schema

    If you are unsure why, please review *Chapter 2, Data Structures, Types, and Formats – Going through the Data Schema and its Types*

20. The answer is: Duplicate data

    If you are unsure why, please review *Chapter 4, Cleaning and Processing Data – Managing Duplicate and Redundant Data*

21. The answer is: Data accuracy

    If you are unsure why, please review *Chapter 15, Data Quality and Management – Understanding Quality Control*

22. The answer is: Cross-validation

    If you are unsure why, please review *Chapter 15, Data Quality and Management – Validating Quality*

23. The answer is: Exploratory data analysis

    If you are unsure why, please review *Chapter 6, Types of Analytics – Exploring Your Data*

24. The answer is: There is a significant difference between Design A and Design B

    If you are unsure why, please review *Chapter 9, Hypothesis Testing – Differentiating Null Hypothesis and Alternative Hypothesis*

25. The answer is: Performance analysis

    If you are unsure why, please review *Chapter 6, Types of Analytics – Checking on Performance*

26. The answer is: 135.5

    If you are unsure why, please review *Chapter 7, Measures of Central Tendency and Dispersion – Finding Variance and Standard Deviation*

27. The answer is: A simple linear regression

    If you are unsure why, please review *Chapter 10, Introduction to Inferential Statistics – Simple Linear Regression*

28. The answer is: Normal

    If you are unsure why, please review *Chapter 7, Measures of Central Tendency and Dispersion – Discovering Distributions*

29. The answer is: Interactive saved searches

    If you are unsure why, please review *Chapter 12, Reporting Process – Understanding Report Delivery*

30. The answer is: JSON

    If you are unsure why, please review *Chapter 2, Data Structures, Types, and Formats – Going through Data Types and File Types*

31. The answer is: 40

    If you are unsure why, please review *Chapter 7, Measures of Central Tendency and Dispersion – Understanding Measures of Central Tendency*

32. The answer is: Data lake

    If you are unsure why, please review *Chapter 2, Data Structures, Types, and Formats – Understanding the Concept of Warehouses and Lakes*

33. The answer is: T-test

    If you are unsure why, please review *Chapter 10, Introduction to Inferential Statistics – Understanding t-Tests*

34. The answer is: Date, Distance(m), and RobotWeight(lb)

    If you are unsure why, please review *Chapter 12, Reporting Process – Knowing what to Consider when Making a Report*

35. The answer is: Chi-square

    If you are unsure why, please review *Chapter 10, Introduction to Inferential Statistics – Knowing Chi-Square*

36. The answer is: Consistency

    If you are unsure why, please review *Chapter 15, Data Quality and Management – Understanding Quality Control*

37. The answer is: A dynamic report

    If you are unsure why, please review *Chapter 11, Types of Reports – Distinguishing Static and Dynamic Reports*

38. The answer is: Defined rows/columns

    If you are unsure why, please review *Chapter 2, Data Structures, Types, and Formats – Understanding Structured and Unstructured Data*

39. The answer is: The reference dates

    If you are unsure why, please review *Chapter 12, Reporting Process – Understanding Report Elements*

40. The answer is: Branding

    If you are unsure why, please review *Chapter 12, Reporting Process – Designing Reports*

41. The answer is: A self-service report

    If you are unsure why, please review *Chapter 11, Types of Reports – Knowing about Self-Service Reports*

42. The answer is: Listwise deletion

     If you are unsure why, please review *Chapter 4, Cleaning and Processing Data – Dealing with Missing Data*

43. The answer is: System functions

     If you are unsure why, please review *Chapter 5, Data Wrangling and Manipulation – Shaping Data with Common Functions*

44. The answer is: Conditional operators

     If you are unsure why, please review *Chapter 5, Data Wrangling and Manipulation – Shaping Data with Common Functions*

45. The answer is: Imputation

     If you are unsure why, please review *Chapter 4, Cleaning and Processing Data – Dealing with Missing Data*

46. The answer is: A stacked bar chart

     If you are unsure why, please review *Chapter 13, Common Visualizations – Comprehending Charts with Bars*

47. The answer is: Data transformation

     If you are unsure why, please review *Chapter 15, Data Quality and Management – Understanding Quality Control*

48. The answer is: Positive correlation

     If you are unsure why, please review *Chapter 10, Introduction to Inferential Statistics – Calculating Correlations*

49. The answer is: Z-score

     If you are unsure why, please review *Chapter 8, Common Techniques in Descriptive Statistics – Understanding z-Scores*

50. The answer is: Relational

     If you are unsure why, please review *Chapter 2, Data Structures, Types, and Formats – Understanding Structured and Unstructured Data*

51. The answer is: Views

     If you are unsure why, please review *Chapter 12, Reporting Process – Knowing what to Consider when Making a Report*

52. The answer is: 47%

    If you are unsure why, please review *Chapter 8, Common Techniques in Descriptive Statistics – Calculating Percent Change and Percent Difference*

53. The answer is: A recurring report

    If you are unsure why, please review *Chapter 11, Types of Reports – Understanding Recurring Reports*

54. The answer is: The same information recorded in multiple columns

    If you are unsure why, please review *Chapter 4, Cleaning and Processing Data – Managing Duplicate and Redundant Data*

55. The answer is: Non-parametric

    If you are unsure why, please review *Chapter 4, Cleaning and Processing Data – Understanding Non-parametric Data*

56. The answer is: PII

    If you are unsure why, please review *Chapter 14, Data Governance – Understanding Data Classifications*

57. The answer is: 0.008

    If you are unsure why, please review *Chapter 9, Hypothesis Testing – Learning p-Value and Alpha*

58. The answer is: Discrete

    If you are unsure why, please review *Chapter 2, Data Structures, Types, and Formats – Going through Data Types and File Types*

59. The answer is: OLTP

    If you are unsure why, please review *Chapter 3, Collecting Data – Understanding OLTP and OLAP*

60. The answer is: Parsing

    If you are unsure why, please review *Chapter 5, Data Wrangling and Manipulation – Parsing Your Data*

61. The answer is: A derived variable

    If you are unsure why, please review *Chapter 5, Data Wrangling and Manipulation – Calculating Derived and Reduced Variables*

62. The answer is: The chi-square goodness of fit

    If you are unsure why, please review *Chapter 10, Introduction to Inferential Statistics – Knowing Chi-Square*

63. The answer is: Link analysis

   If you are unsure why, please review *Chapter 6, Types of Analytics – Finding Links*

64. The answer is: A line chart

   If you are unsure why, please review *Chapter 13, Common Visualizations – Charting Lines, Circles, and Dots*

65. The answer is: The cover page

   If you are unsure why, please review *Chapter 12, Reporting Process – Understanding Report Elements*

66. The answer is: There are at least 60 employees in the company

   If you are unsure why, please review *Chapter 13, Common Visualizations – Comprehending Charts with Bars*

67. The answer is: A bubble chart

   If you are unsure why, please review *Chapter 13, Common Visualizations – Charting Lines, Circles, and Dots*

68. The answer is: An outlier

   If you are unsure why, please review *Chapter 4, Cleaning and Processing Data – Finding Outliers*

69. The answer is: The acceptable use policy

   If you are unsure why, please review *Chapter 14, Data Governance – Knowing Use Requirements*

70. The answer is: Inner join

   If you are unsure why, please review *Chapter 5, Data Wrangling and Manipulation – Merging Data*

71. The answer is: When a company is purchased

   If you are unsure why, please review *Chapter 15, Data Quality and Management – Understanding Master Data Management (MDM)*

72. The answer is: Active Record

   If you are unsure why, please review *Chapter 2, Data Structures, Types, and Formats – Updating Stored Data*

73. The answer is: A data warehouse

   If you are unsure why, please review *Chapter 2, Data Structures, Types, and Formats – Understanding the Concept of Warehouses and Lakes*

74. The answer is: Dummy coding

   If you are unsure why, please review *Chapter 5, Data Wrangling and Manipulation – Recoding Variables*

75. The answer is: 25

   If you are unsure why, please review *Chapter 7, Measures of Central Tendency and Dispersion – Calculating Range and Quartiles*

76. The answer is: A star schema

   If you are unsure why, please review *Chapter 2, Data Structures, Types, and Formats – Going through the Data Schema and its Types*

77. The answer is: Role-based

   If you are unsure why, please review *Chapter 14, Data Governance – Understanding Data Security*

78. The answer is: Reject the alternative hypothesis and accept the null hypothesis

   If you are unsure why, please review *Chapter 9, Hypothesis Testing – Learning p-Value and Alpha*

79. The answer is: Parameterization

   If you are unsure why, please review *Chapter 3, Collecting Data – Optimizing Query Structure*

80. The answer is: Data type validation

   If you are unsure why, please review *Chapter 4, Cleaning and Processing Data – Understanding Invalid Data, Specification Mismatch, and Data Type Validation*

81. The answer is: Synchronous

   If you are unsure why, please review *Chapter 3, Collecting Data – Utilizing Public Sources of Data*

82. The answer is: One-to-many

   If you are unsure why, please review *Chapter 14, Data Governance – Handling Entity Relationship Requirements*

83. The answer is: Data audits

   If you are unsure why, please review *Chapter 15, Data Quality and Management – Validating Quality*

84. The answer is: The taller you are the more likely you are to have a larger hat size

   If you are unsure why, please review *Chapter 13, Common Visualizations – Charting Lines, Circles, and Dots*

85. The answer is: Audio

   If you are unsure why, please review *Chapter 2, Data Structures, Types, and Formats – Going through Data Types and File Types*

86. The answer is: Web scraping

   If you are unsure why, please review *Chapter 3, Collecting Data – Collecting Your Own Data*

87. The answer is: Both structured and relational

    If you are unsure why, please review *Chapter 2, Data Structures, Types, and Formats – Understanding Structured and Unstructured Data*

88. The answer is: Data dictionary

    If you are unsure why, please review *Chapter 15, Data Quality and Management – Understanding Master Data Management (MDM)*

89. The answer is: Sorting

    If you are unsure why, please review *Chapter 3, Collecting Data – Optimizing Query Structure*

90. The answer is: Type I

    If you are unsure why, please review *Chapter 9, Hypothesis Testing – Understanding Type I and Type II Error*

17
Practice Exam Two

If you are here, you have already taken the first practice exam and reviewed the content you missed. You know the drill!

To make this exam the best practice it can be, you should follow the testing experience as closely as possible. Make sure you are in a quiet and clean environment and have access to a simple calculator, a timer, and the ability to take notes.

Set your timer for 90 minutes.

Don't panic.

Take a deep breath.

You can do this.

Practice exam two

1. EDA stands for what?

    A. Extra data analysis

    B. Extrapolate data architecture

    C. Extract data analyze

    D. Exploratory data analysis

2. You run a study to see whether there is a difference in the amount of weight a hamster can lift compared to a gerbil. You find out the results and come to a conclusion. If the result was a type II error, what was your conclusion?

    A. There is a difference between hamster and gerbil lifting capacities

    B. Gerbils and hamsters can lift the same amount

    C. Hamsters can lift more than gerbils

    D. Gerbils can lift more than hamsters

3. The following dataset is an example of what type of error?

ID	Sex	Male	Female
84927	M	TRUE	FALSE
69427	M	TRUE	FALSE
10374	F	FALSE	TRUE
58264	M	TRUE	FALSE
90162	F	FALSE	TRUE

    A. Missing data

    B. Duplicate data

    C. Invalid data

    D. Redundant data

4. Which of the following represents the percent of observations in each category as compared to the whole?

    A. Percentage

    B. Percent change

    C. Percent difference

    D. Frequency

5. A pharmaceutical lab wants to know whether a memory-enhancing drug works. They have two mice run the same maze multiple times, one with the medication and one without. You analyze the results and receive a p-value of 0.04. Assuming an alpha of 0.05, how do you interpret the results?

    A. Accept the alternative hypothesis and reject the null hypothesis

    B. Reject the alternative hypothesis and accept the null hypothesis

    C. Accept the alternative and null hypotheses

    D. Reject the alternative and null hypotheses

6. A company that produces cleaning supplies has a new product. To see whether it is effective, they run a trial with two groups. The first group uses the new product to clean a stain and the second group uses tap water to clean a stain. The idea that there will be no difference between the performance of these two groups is what kind of hypothesis?

    A. Alternative hypothesis

    B. Null hypothesis

    C. Secondary hypothesis

    D. Original hypothesis

7. A small ferret farm would like to see whether there is a relationship between the weight of a ferret in grams and how much milk it produces in milliliters. Which of the following visualizations would be most appropriate?

    A. A line chart

    B. A bubble chart

    C. A scatter plot

    D. A tree map

8. A flat file delimited by commas is what file type?

    A. JPEG

    B. TSV

    C. AAC

    D. CSV

9. Which of the following elements should never be on the cover page of a report?

    A. The version number

    B. The report run data

    C. The data refresh date

    D. The appendix

10. Data type validation is a process specifically used to avoid what type of error?

    A. A specification mismatch

    B. Invalid data

    C. Missing data

    D. Duplicate data

11. What is an appropriate title for the following chart?

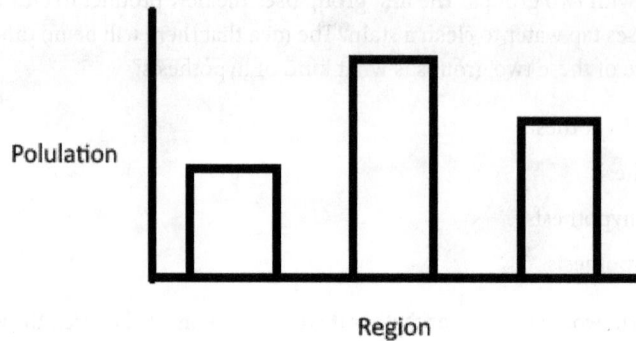

Polulation

Region

A. The Population of India based on the Geographical Region

B. Population

C. The Population of India Averaged for the Years 2015, 2016, and 2017 as Sub-Divided by Geographical Regions Determined by the 2018 Land Survey

D. Region 2010

12. A mortgage company would like its sales representatives to have access to a dashboard with the absolute most up-to-date rates and figures. This means that the dashboard should be what?

A. Heavily filtered

B. Subscription-based

C. Real-time

D. Point-in-time

13. The act of automatically moving and analyzing online transactions is called what?

A. OLTP

B. OLAP

C. ELT

D. ETL

14. The following code snippet is an example of what concept?

```
Data = "This book makes me happy."
Data = ["This", "book", "makes", "me", "happy", "."]
```

A. Reduction

B. Interpretation

C. Imputation

D. Parsing

15. The following chart represents what kind of visualization?

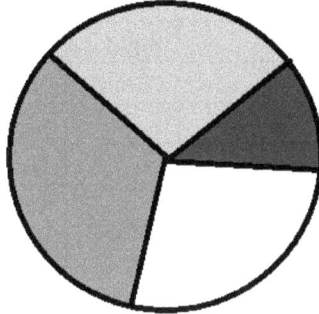

A. A pie chart

B. A scatter plot

C. A histogram

D. A heat map

16. Which of the following is a valid data storage solution for audio files?

A. A data mart

B. A data lake

C. A data warehouse

D. A data store

17. An engineer would like to check how efficient each phase of a production process is to see whether it can be improved. What type of analysis is most appropriate?

A. Performance analysis

B. Link analysis

C. Trend analysis

D. Exploratory data analysis

18. You generate a highly detailed report on the number of eggs every single chicken a company owns produces compared to how much that chicken has eaten to get a specific grain-to-egg efficiency ratio for each animal. Who is the most appropriate audience for this report?

A. C-level executives

B. Stakeholders

C. The general public

D. Technical experts

19. If you were to perform an outer join on the following tables, what would be the result?

Left Table		Right Table	
ClientID	Name	ClientID	Name
1	Smith, Laurence	1	Austin, TX
2	Brown, Betty	2	Denver, CO
3	Hook, Phil	3	Tulsa, OK
4	Roark, Jen	7	Phoenix, AZ
5	Cox, Jona	8	Seattle, WA
6	Humbert, Ren	9	Baltimore, MD

A.

Joined Table		
ClientID	Name	City
1	Smith, Laurence	Austin, TX
2	Brown, Betty	Denver, CO
3	Hook, Phil	Tulsa, OK
4	Roark, Jen	NULL
5	Cox, Jona	NULL
6	Humbert, Ren	NULL
7	NULL	Phoenix, AZ
8	NULL	Seattle, WA
9	NULL	Baltimore, MD

B.

Joined Table		
ClientID	Name	City
1	Smith, Laurence	Austin, TX
2	Brown, Betty	Denver, CO
3	Hook, Phil	Tulsa, OK

C.

Joined Table		
ClientID	Name	City
1	Smith, Laurence	Austin, TX
2	Brown, Betty	Denver, CO
3	Hook, Phil	Tulsa, OK
4	Roark, Jen	NULL
5	Cox, Jona	NULL
6	Humbert, Ren	NULL

D.

Joined Table		
ClientID	Name	City
1	Smith, Laurence	Austin, TX
2	Brown, Betty	Denver, CO
3	Hook, Phil	Tulsa, OK
7	NULL	Phoenix, AZ
8	NULL	Seattle, WA
9	NULL	Baltimore, MD

20. A project manager would like an operational report at the end of every sprint. What type of report would be most appropriate?

   A. A research report

   B. An ad hoc report

   C. A self-service report

   D. A recurring report

21. Find the mode of the following dataset:

$$5, 3, 8, 5, 3, 9, 3, 8, 2$$

   A. 2

   B. 3

   C. 5

   D. 9

22. A project manager wants to know whether there is a connection between how many hours their team works in a day and how many mistakes they make. What type of analysis is most appropriate?

    A. Link analysis

    B. Trend analysis

    C. Performance analysis

    D. Exploratory data analysis

23. The following table is in chronological order, with new values added to the bottom. This table is an example of updating a table by what means?

Total Number of Beans	Red Beans	Blue Beans	Yellow Beans
10	3	4	3
12	2	6	4
11	X	X	X
9	X	X	X
10	X	X	X

    A. Adding variables

    B. Removing variables

    C. Deleting historical data

    D. Active record

24. What conclusion can you draw from the following visualization?

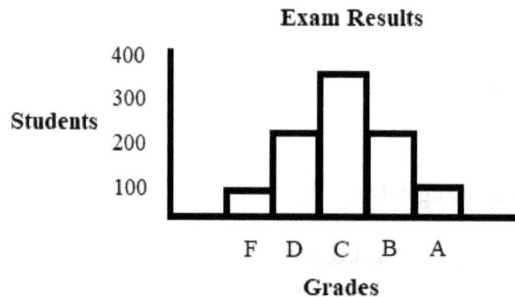

Exam Results

    A. The majority of students failed the exam

    B. The distribution is nonparametric

    C. Around 350 students achieved a grade of C or higher

    D. A new student taking the test would most likely get a C

25. The following chart depicts what kind of visualization?

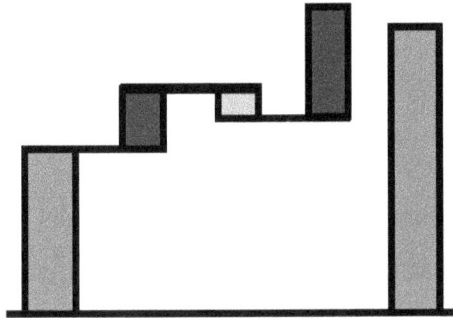

A.   A waterfall chart

B.   A histogram

C.   A stacked bar chart

D.   A line chart

26. A schema with only normalized tables would be what type of schema?

A.   A galaxy schema

B.   A star schema

C.   A fast constellation schema

D.   A snowflake schema

27. The following dataset is an example of what concept?

Month	UnitsSold	Color	ColorRecoded
August	432	Red	1
August	365	Blue	2
August	154	Yellow	3
September	398	Red	1
September	386	Blue	2
September	108	Yellow	3

A.   Recoding a category into a number

B.   Recoding a number into a category

C.   Transposition

D.   Dummy coding

28. A detailed program that explains how the software performs a specific query is called what?

    A. An execution plan

    B. A subquery

    C. Parameterization

    D. A temporary table

29. What conclusion can you draw from the following visualization?

Job Titles Who Have Access to Company Data

A. If you were to pick a person who can access data at random, they would most likely be a business analyst

B. You need to have "analyst" in your title to access data

C. The other group is not as technically skilled

D. Half of everyone who can access data is either a marketing analyst or a business analyst

30. A small, highly specialized data storage solution following a star schema would most likely be what?

    A. A data warehouse

    B. A data lake

    C. A data pond

    D. A data mart

31. The CEO of your company is considering a merger with your company's main competitor. They would like a detailed report on the pros and cons, as well as projections on all of the major KPIs for the next 10 years. This report is due in 6 months. What type of report is most appropriate in this scenario?

    A. A self-service report

    B. A research report

    C. A recurring report

    D. An ad hoc report

32. Which type of schema has two levels of dimension tables?

    A. A star schema

    B. A snowball schema

    C. A snowflake schema

    D. A galaxy schema

33. In the following table, which variable indicates when a variable stopped being active?

Magic Number	Active Record	Active Start	Active End
41	No	11/11/2011	12/12/2012
42	Yes	12/12/2012	

    A. Magic Number

    B. Active Record

    C. Active Start

    D. Active End

34. Which of the following is a key process of MDM?

    A. Data manipulation

    B. Data encryption

    C. Data transformation

    D. Data consolidation

35. The following table is a sample from a larger dataset. What type of visualization would be most appropriate to display this information?

Population of Europe by Country	
Country	Population (million)
Russia	145
Germany	84
United Kingdom	68
France	65
Italy	60
Spain	47
...	...

   A. A waterfall chart

   B. A geographic map

   C. A pie chart

   D. A heat map

36. The following dataset is an example of what type of error?

Cost Per Click
$1.82
$0.95
Toast
$2.10
$1.39

   A. Invalid data

   B. Specification mismatch

   C. Redundant data

   D. Data type validation

37. The following screenshot represents what type of survey question?

   2. What is your favorite kind of peanut butter? ♀ 0

   ○ Smooth

   ○ Chunky

A.   Text-based

B.   Single choice

C.   Multiple choice

D.   Drop-down

38. Which of the following is a conditional operator?

A.   ORDER BY

B.   APPEND

C.   SUM

D.   OR

39. Which of the following is something to consider when checking for data quality?

A.   Data visualization

B.   Data transmission

C.   Data manipulation

D.   Data integrity

40. What data-validating approach should you take if you see the results of an analysis and you believe them to be in error?

A.   Data audits

B.   Data profiling

C.   Reasonable expectations

D.   Cross-validation

41. Find the standard deviation of the following dataset:

62, 92, 43, 66, 37

A.   21.7

B.   34.6

C.   60.0

D.   470.5

42. If you are using natural language processing to analyze a text file and would like to visualize a way to express the ideas held within the text, what is the most appropriate visualization?

    A. A word cloud

    B. A tree map

    C. A waterfall chart

    D. A line chart

43. When you suspect that private data might have been breached, what is the single most important thing you can do?

    A. Notify the impacted parties

    B. Fix the breach

    C. Ignore the breach

    D. Figure out how the breach happened

44. The following dataset is an example of what type of data?

Total Number of Beans	Red Beans	Blue Beans	Yellow Beans
10	3	4	3
12	2	6	4
11	3	5	3

    A. Structured

    B. Unstructured

    C. Semi-structured

    D. Non-relational

45. When you are creating a dashboard, what should you do immediately after planning out your data story?

    A. Deliver the dashboard

    B. Create the dashboard

    C. Get approval

    D. Plan a second data story

46. You run a simple A/B study to compare two different advertising campaigns. Assuming an alpha of 0.1, which of the following p-values would cause you to accept the null hypothesis?

    A. 0.05

    B. 0.09

    C. 0.3

    D. 0.001

47. The marketing department has created a customer persona, and they would like to compare the age of their persona against the normally distributed ages of their actual customers. Which analysis is most appropriate?

    A. Chi-square

    B. Simple linear regression

    C. Z-score

    D. T-test

48. The following tables have what sort of cardinality?

Employee Information		Employee Departments	
EmployeeID	EmployeeName	EmployeeID	Department
8279	Steve	8279	Sales
7904	Joan	7904	Human Resources
2905	Frank	2905	Marketing
8391	Bertha	8391	Administration

    A. One-to-one

    B. One-to-many

    C. Many-to-many

    D. There is no entity relationship

49. If you had a variable called BirdPassed that kept track of whether or not a bird passed by your window on a given day with YES or NO, what variable type would it be?

    A. Ordinal

    B. Binary

    C. Discrete

    D. Continuous

50. You calculate that the average number of clicks per minute for the year for a small e-commerce website is 12.8. Another website that is a major competitor has an average number of clicks per minute for the year of 13.7. What is the difference between your value and your competitors?

   A. 7%

   B. 9%

   C. 12%

   D. 18%

51. Given a t-value of 1.86, which is the confidence interval for the following dataset?

   8, 7, 8, 8, 10, 6, 8, 8, 9, 8

   A. 7.3 to 8.7

   B. 6.9 to 9.1

   C. 8.4 to 8.8

   D. 7.6 to 9.5

52. The following is an example of what type of join?

Left Table		Joined Table			Right Table	
ClientID	Name	ClientID	Name	City	ClientID	City
1	Smith, Laurence	1	Smith, Laurence	Austin, TX	1	Austin, TX
2	Brown, Betty	2	Brown, Betty	Denver, CO	2	Denver, CO
3	Hook, Phil	3	Hook, Phil	Tulsa, OK	3	Tulsa, OK
4	Roark, Jen	4	Roark, Jen	NULL	7	Phoenix, AZ
5	Cox, Jona	5	Cox, Jona	NULL	8	Seattle, WA
6	Humbert, Ren	6	Humbert, Ren	NULL	9	Baltimore, MD

   A. Outer join

   B. Inner Join

   C. Left Join

   D. Right Join

53. An analysis that specifically tells you whether or not two categorical variables are related best describes what analysis?

   A. Z-score

   B. Chi-square

   C. T-test

   D. Simple linear regression

54. A social media ID is considered what type of protected data?

    A. PII
    B. PHI
    C. PCI
    D. PIFI

55. What is duplicate data?

    A. Data that does not meet formatting requirements
    B. The same information recorded in multiple columns
    C. Data that is incomplete or blank
    D. The same information recorded in multiple rows

56. The following is an example of what type of error?

    A. Specification mismatch
    B. An outlier
    C. Redundant data
    D. Missing data

57. Which of the following is considered a public source of data?

    A. Surveying
    B. Web services
    C. Web scraping
    D. Studies

58. Average, sum, and count are all examples of what?

    A. Augmentation

    B. Conditional operators

    C. Reduction

    D. Parsing

59. Which of the following are circumstances under which you should check the quality of your data?

    A. Data encryption

    B. Data transmission

    C. Data acquisition

    D. Data deletion

60. What data-validating approach should you take if you need a formal process to apply to an entire database?

    A. Data audits

    B. Data profiling

    C. Spot checking

    D. Cross-validation

61. You are given a dataset that includes tree width and how much weight it can hold before it starts to bend. Then, you are asked to predict how wide a tree must be before it can hold a 400 lb sumo wrestler without bending. Which analysis is most appropriate?

    A. Simple linear regression

    B. T-test

    C. Chi-square test for independence

    D. Chi-square goodness of fit

62. The following dataset is an example of what concept?

Month	UnitsSold	Color
August	432	Red
August	365	Blue
August	154	Yellow
September	398	Red
September	386	Blue
September	108	Yellow

Month	August	August	August	September	September	September
UnitsSold	432	365	154	398	386	108
Color	Red	Blue	Yellow	Red	Blue	Yellow

A. Dummy coding

B. Transposition

C. Reduction

D. Conditional operators

63. Making sure your data is not full of gaps and missing data is considered what data quality dimension?

A. Accuracy

B. Completeness

C. Consistency

D. Attribute limitation

64. The following chart represents what type of distribution?

A. Normal

B. Uniform

C. Exponential

D. Poisson

65. A person's medical record is considered what type of protected data?

A. PII

B. PHI

C. PCI

D. PIFI

66. In general, dashboards are considered what type of report?

A. A self-service report

B. A risk and regulatory report

C. An ad hoc report

D. A research report

67. Your manager gives you access to historical data for a certain variable and would like you to predict what might happen with that variable in the future. What type of analysis is most appropriate?

   A.   Link analysis

   B.   Trend analysis

   C.   Performance analysis

   D.   Exploratory data analysis

68. Which of the following analytical tools is specialized for visualizations?

   A.   AWS QuickSight

   B.   SQL

   C.   Apex Systems

   D.   Stata

69. A line manager in a production plant would like to know the specific efficiencies of every machine to see which need tuning at the end of the week. What is the most appropriate data range for this report?

   A.   Weeks

   B.   Months

   C.   Years

   D.   Decades

70. Deleting only the missing values and only as they are needed is what type of deletion?

   A.   Pairwise deletion

   B.   Listwise deletion

   C.   Variable deletion

   D.   Filtering

71. Your manager has requested a one-time report to answer a specific business question. What type of report is most appropriate in this scenario?

   A.   A static report

   B.   A recurring report

   C.   A dashboard

   D.   A dynamic report

72. After publishing a dashboard, you continue to receive emails asking the same questions over and over again. What part of the dashboard should you update to save time?

    A. The cover page

    B. The FAQs

    C. The appendix

    D. The reference data sources

73. Unstructured databases include which of the following data types?

    A. Undefined fields

    B. Machine data

    C. Undefined fields and machine data

    D. Neither undefined fields nor machine data

74. Which of the following file types can be used to structure a website or pass data through a website?

    A. WMA

    B. XML

    C. AVI

    D. WMV

75. Find the middle quartile (Q2) of the following dataset:

    70, 21, 34, 48, 27

    A. 21

    B. 34

    C. 40

    D. 70

76. Watching things and taking notes as a form of data collection is called what?

    A. API

    B. Web scraping

    C. Survey

    D. Observation

77. The following dataset depicts an action that has been performed on a dataset. What is the name of that action?

Employee ID	LastName	FirstName	Department	YearsWithCompany
83784	Benhill	Floyd	Sales	12
64986	Chane	Jill	IT	1
93671	Hanson	Richard	HR	15
37816	Smith	Trudy	Sales	21

Employee ID	LastName	FirstName	Department	YearsWithCompany
83784	Benhill	Floyd	Sales	12
37816	Smith	Trudy	Sales	21

A. Filtering

B. Indexing

C. Sorting

D. Execution planning

78. What do you call the process of filling gaps in the data by calculating the most likely value based on the values of other variables in the row?

A. Interpolation

B. Interpretation

C. Imputation

D. Infusion

79. How do nonparametric distributions relate to normal distributions?

A. Nonparametric distributions are never normal

B. Nonparametric distributions are sometimes normal

C. Normal distributions are sometimes parametric

D. Normal distributions are always nonparametric

80. The HR department at your company wants to know whether or not there is a relationship between an employee's job title and the color of their hair. You don't know why they think this is important, but what type of analysis would be most appropriate here?

   A.   Chi-square test for independence

   B.   Chi-square test for homogeneity

   C.   Chi-square goodness of fit

   D.   Chi-square test for linearity

81. A delta load happens when you do what?

   A.   Load information into a new location for the first time

   B.   Upload all information

   C.   Only load information that is new or has changed

   D.   Only load information that hasn't changed

82. The following diagram represents what type of database schema?

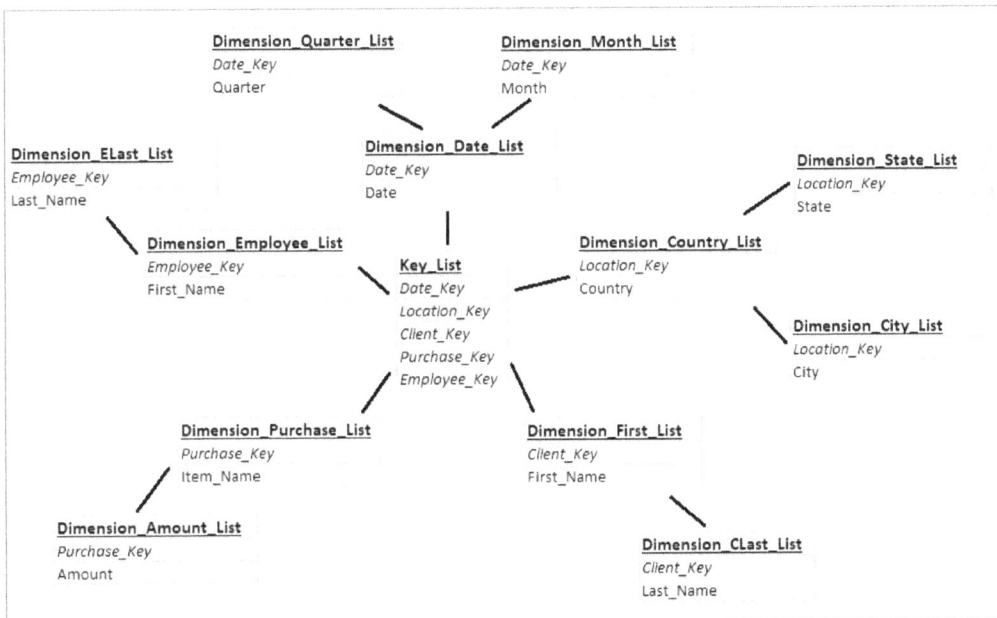

   A.   A snowflake schema

   B.   A star schema

   C.   A galaxy schema

   D.   A fast constellation schema

83. A company fundraiser would like to know how many shirts were sold of each shirt size so that they can create a distribution and order the appropriate numbers for the next one. What type of visualization would be most appropriate?

    A. A histogram
    B. A waterfall chart
    C. A stacked bar chart
    D. A heat map

84. Which of the following would you find in a structured database?

    A. Machine data
    B. Text files
    C. Emails
    D. Key-value pairs

85. An analysis that compares quantitative variables to see whether there is a relationship between them and how strong that relationship is best describes what analysis?

    A. Z-score
    B. Chi-square
    C. Correlation
    D. T-test

86. You are required by law to make certain data illegible during transit by translating the data from plaintext to cyphertext. This data cannot be accessed without a specific key. What security process is described here?

    A. Release approval
    B. Data masking
    C. Access requirements
    D. Data encryption

87. An analysis that compares two groups of quantitative variables to tell whether or not there is a significant difference between them most accurately describes what type of analysis?

    A. Correlation
    B. Z-score
    C. T-test
    D. Chi-square

88. What is a major benefit of MDM?

    A. Streamlining data access

    B. Creating a diversified database

    C. Requiring more joins

    D. Safely storing the data in multiple places

89. Your boss would like you to create a dashboard that automatically refreshes and sends out invitations to the appropriate parties every Monday morning. What is the most suitable approach?

    A. Access permissions

    B. Subscription

    C. Interactive saved searches

    D. Scheduled delivery

90. Which section of the data use agreement includes information on when and how data will be destroyed?

    A. The acceptable use policy

    B. Data processing

    C. Data deletion

    D. Data retention

Congratulations!

You made it through your second practice exam! That wasn't so bad, right? Take a moment to celebrate how far you've come. When you are ready, compare your answers to the answers listed here. Again, your goal is to get fewer than 20 questions wrong as a rough guideline. Make sure to review the topics of the questions you missed.

When you are ready, you can go to the following website, where you can schedule your exam or find additional resources if you are still struggling: `https://www.comptia.org/certifications/data`.

This wraps up every topic on the exam, as well as over 200 practice questions! You are ready.

Thank you for reading this book.

Best of luck with your exam!

Take a deep breath.

You can do it.

Practice exam two answers

1. The answer is: Exploratory data analysis

   If you are unsure why, please review *Chapter 6, Types of Analytics – Exploring Your Data*

2. The answer is: Gerbils and hamsters can lift the same amount

   If you are unsure why, please review *Chapter 9, Hypothesis Testing – Understanding Type I and Type II Error*

3. The answer is: Redundant data

   If you are unsure why, please review *Chapter 4, Cleaning and Processing Data – Managing Duplicate and Redundant Data*

4. The answer is: Percentage

   If you are unsure why, please review *Chapter 8, Common Techniques in Descriptive Statistics – Understanding Frequencies and Percentages*

5. The answer is: Accept the alternative hypothesis and reject the null hypothesis

   If you are unsure why, please review *Chapter 9, Hypothesis Testing – Learning P-Value and Alpha*

6. The answer is: Null hypothesis

   If you are unsure why, please review *Chapter 9, Hypothesis Testing – Differentiating Null Hypothesis and Alternative Hypothesis*

7. The answer is: A scatter plot

   If you are unsure why, please review *Chapter 13, Common Visualizations – Charting Lines, Circles, and Dots*

8. The answer is: CSV

   If you are unsure why, please review *Chapter 2, Data Structures, Types, and Formats – Going Through Data Types and File Types*

9. The answer is: The appendix

   If you are unsure why, please review *Chapter 12, Reporting Process – Understanding Report Elements*

10. The answer is: A specification mismatch

    If you are unsure why, please review *Chapter 4, Cleaning and Processing Data – Understanding Invalid Data, Specification Mismatch, and Data Type Validation*

11. The answer is: The Population of India based on the Geographical Region

    If you are unsure why, please review *Chapter 12, Reporting Process – Designing Reports*

12. The answer is: Real-time

    If you are unsure why, please review *Chapter 12, Reporting Process – Understanding Report Delivery*

13. The answer is: OLAP

    If you are unsure why, please review *Chapter 3, Collecting Data – Understanding OLTP and OLAP*

14. The answer is: Parsing

    If you are unsure why, please review *Chapter 5, Data Wrangling and Manipulation – Parsing Your Data*

15. The answer is: A pie chart

    If you are unsure why, please review *Chapter 13, Common Visualizations – Charting Lines, Circles, and Dots*

16. The answer is: A data lake

    If you are unsure why, please review *Chapter 2, Data Structures, Types, and Formats – Understanding the Concept of Warehouses and Lakes*

17. The answer is: Performance analysis

    If you are unsure why, please review *Chapter 6, Types of Analytics – Checking on Performance*

18. The answer is: Technical experts

    If you are unsure why, please review *Chapter 12, Reporting Process – Knowing what to Consider when Making a Report*

19. The answer is:

Joined Table		
ClientID	Name	City
1	Smith, Laurence	Austin, TX
2	Brown, Betty	Denver, CO
3	Hook, Phil	Tulsa, OK
4	Roark, Jen	NULL
5	Cox, Jona	NULL
6	Humbert, Ren	NULL
7	NULL	Phoenix, AZ
8	NULL	Seattle, WA
9	NULL	Baltimore, MD

    If you are unsure why, please review *Chapter 5, Data Wrangling and Manipulation – Merging Data*

20. The answer is: A recurring report

   If you are unsure why, please review *Chapter 11, Types of Reports – Understanding Recurring Reports*

21. The answer is: 3

   If you are unsure why, please review *Chapter 7, Measures of Central Tendency and Dispersion – Understanding Measures of Central Tendency*

22. The answer is: Link analysis

   If you are unsure why, please review *Chapter 6, Types of Analytics – Finding Links*

23. The answer is: Removing variables

   If you are unsure why, please review *Chapter 2, Data Structures, Types, and Formats – Updating Stored Data*

24. The answer is: A new student taking the test would most likely get a C

   If you are unsure why, please review *Chapter 13, Common Visualizations – Comprehending Charts with Bars*

25. The answer is: A waterfall chart

   If you are unsure why, please review *Chapter 13, Common Visualizations – Comprehending Charts with Bars*

26. The answer is: A snowflake schema

   If you are unsure why, please review *Chapter 2, Data Structures, Types, and Formats – Going Through the Data Schema and its Types*

27. The answer is: Recoding a category into a number

   If you are unsure why, please review *Chapter 5, Data Wrangling and Manipulation – Recoding Variables*

28. The answer is: An execution plan

   If you are unsure why, please review *Chapter 3, Collecting Data – Optimizing Query Structure*

29. The answer is: Half of everyone who can access data is either a marketing analyst or a business analyst

   If you are unsure why, please review *Chapter 13, Common Visualizations – Charting Lines, Circles, and Dots*

30. The answer is: A data mart

   If you are unsure why, please review *Chapter 2, Data Structures, Types, and Formats – Understanding the Concept of Warehouses and Lakes*

31. The answer is: A research report

    If you are unsure why, please review *Chapter 11, Types of Reports – Understanding Ad hoc and Research Reports*

32. The answer is: A snowflake schema

    If you are unsure why, please review *Chapter 2, Data Structures, Types, and Formats – Going Through the Data Schema and its Types*

33. The answer is: Active End

    If you are unsure why, please review *Chapter 2, Data Structures, Types, and Formats – Updating Stored Data*

34. The answer is: Data consolidation

    If you are unsure why, please review *Chapter 15, Data Quality and Management – Understanding Master Data Management (MDM)*

35. The answer is: A geographic map

    If you are unsure why, please review *Chapter 13, Common Visualizations – Understanding Heat Maps, Tree Maps, and Geographic Maps*

36. The answer is: Specification mismatch

    If you are unsure why, please review *Chapter 4, Cleaning and Processing Data – Understanding Invalid Data, Specification Mismatch, and Data Type Validation*

37. The answer is: Single choice

    If you are unsure why, please review *Chapter 3, Collecting Data – Collecting Your Own Data*

38. The answer is: OR

    If you are unsure why, please review *Chapter 5, Data Wrangling and Manipulation – Shaping Data with Common Functions*

39. The answer is: Data integrity

    If you are unsure why, please review *Chapter 15, Data Quality and Management – Understanding Quality Control*

40. The answer is: Reasonable expectations

    If you are unsure why, please review *Chapter 15, Data Quality and Management – Validating Quality*

41. The answer is: 21.7

    If you are unsure why, please review *Chapter 7, Measures of Central Tendency and Dispersion – Finding Variance and Standard Deviation*

42. The answer is: A word cloud

    If you are unsure why, please review *Chapter 13, Common Visualizations – Understanding Infographics and Word Clouds*

43. The answer is: Notify the impacted parties

    If you are unsure why, please review *Chapter 14, Data Governance – Knowing Use Requirements*

44. The answer is: Structured

    If you are unsure why, please review *Chapter 2, Data Structures, Types, and Formats – Understanding Structured and Unstructured Data*

45. The answer is: Get approval

    If you are unsure why, please review *Chapter 12, Reporting Process – Understanding the Report Development Process*

46. The answer is: 0.3

    If you are unsure why, please review *Chapter 9, Hypothesis Testing – Learning p-Value and Alpha*

47. The answer is: Z-score

    If you are unsure why, please review *Chapter 8, Common Techniques in Descriptive Statistics – Understanding Z-Scores*

48. The answer is: One-to-one

    If you are unsure why, please review *Chapter 14, Data Governance – Handling Entity Relationship Requirements*

49. The answer is: Binary

    If you are unsure why, please review *Chapter 2, Data Structures, Types, and Formats – Going Through Data Types and File Types*

50. The answer is: 7%

    If you are unsure why, please review *Chapter 8, Common Techniques in Descriptive Statistics – Calculating Percent Change and Percent Difference*

51. The answer is: 7.3 to 8.7

    If you are unsure why, please review *Chapter 8, Common Techniques in Descriptive Statistics – Discovering Confidence Intervals*

52. The answer is: Left join

    If you are unsure why, please review *Chapter 5, Data Wrangling and Manipulation – Merging Data*

53. The answer is: Chi-square

    If you are unsure why, please review *Chapter 10, Introduction to Inferential Statistics – Knowing Chi-Square*

54. The answer is: PII

    If you are unsure why, please review *Chapter 14, Data Governance – Understanding Data Classifications*

55. The answer is: The same information recorded in multiple rows

    If you are unsure why, please review *Chapter 4, Cleaning and Processing Data – Managing Duplicate and Redundant Data*

56. The answer is: An outlier

    If you are unsure why, please review *Chapter 4, Cleaning and Processing Data – Finding Outliers*

57. The answer is: Web services

    If you are unsure why, please review *Chapter 3, Collecting Data – Utilizing Public Sources of Data*

58. The answer is: Reduction

    If you are unsure why, please review *Chapter 5, Data Wrangling and Manipulation – Calculating Derived and Reduced Variables*

59. The answer is: Data acquisition

    If you are unsure why, please review *Chapter 15, Data Quality and Management – Understanding Quality Control*

60. The answer is: Data profiling

    If you are unsure why, please review *Chapter 15, Data Quality and Management – Validating Quality*

61. The answer is: Simple linear regression

    If you are unsure why, please review *Chapter 10, Introduction to Inferential Statistics – Simple Linear Regression*

62. The answer is: Transposition

    If you are unsure why, please review *Chapter 5, Data Wrangling and Manipulation – Shaping Data with Common Functions*

63. The answer is: Completeness

    If you are unsure why, please review *Chapter 15, Data Quality and Management – Understanding Quality Control*

64. The answer is: Uniform

    If you are unsure why, please review *Chapter 7, Measures of Central Tendency and Dispersion – Discovering Distributions*

65. The answer is: PHI

    If you are unsure why, please review *Chapter 14, Data Governance – Understanding Data Classifications*

66. The answer is: A self-service report

    If you are unsure why, please review *Chapter 11, Types of Reports – Knowing about Self-Service Reports*

67. The answer is: Trend analysis

    If you are unsure why, please review *Chapter 6, Types of Analytics – Discovering Trends*

68. The answer is: AWS QuickSight

    If you are unsure why, please review *Chapter 11, Types of Reports – Knowing Important Analytical Tools*

69. The answer is: Weeks

    If you are unsure why, please review *Chapter 12, Reporting Process – Knowing What to Consider When Making a Report*

70. The answer is: Pairwise deletion

    If you are unsure why, please review *Chapter 4, Cleaning and Processing Data – Dealing with Missing Data*

71. The answer is: A static report

    If you are unsure why, please review *Chapter 11, Types of Reports – Distinguishing Static and Dynamic Reports*

72. The answer is: The FAQs

    If you are unsure why, please review *Chapter 12, Reporting Process – Understanding Report Elements*

73. The answer is: Undefined fields and machine data

    If you are unsure why, please review *Chapter 2, Data Structures, Types, and Formats – Understanding Structured and Unstructured Data*

74. The answer is: XML

    If you are unsure why, please review *Chapter 2, Data Structures, Types, and Formats – Going Through Data Types and File Types*

75. The answer is: 34

    If you are unsure why, please review *Chapter 7, Measures of Central Tendency and Dispersion – Calculating Range and Quartiles*

76. The answer is: Observation

    If you are unsure why, please review *Chapter 3, Collecting Data – Collecting Your Own Data*

77. The answer is: Filtering

    If you are unsure why, please review *Chapter 3, Collecting Data – Optimizing Query Structure*

78. The answer is: Interpolation

    If you are unsure why, please review *Chapter 4, Cleaning and Processing Data – Dealing with Missing Data*

79. The answer is: Nonparametric distributions are never normal

    If you are unsure why, please review *Chapter 4, Cleaning and Processing Data – Understanding Non-Parametric Data*

80. The answer is: Chi-square test for independence

    If you are unsure why, please review *Chapter 10, Introduction to Inferential Statistics – Knowing Chi-Square*

81. The answer is: Only load information that is new or has changed

    If you are unsure why, please review *Chapter 3, Collecting Data – Differentiating ETL and ELT*

82. The answer is: A snowflake schema

    If you are unsure why, please review *Chapter 2, Data Structures, Types, and Formats – Going Through the Data Schema and its Types*

83. The answer is: A histogram

    If you are unsure why, please review *Chapter 13, Common Visualizations – Comprehending Charts with Bars*

84. The answer is: Key-value pairs

    If you are unsure why, please review *Chapter 2, Data Structures, Types, and Formats – Understanding Structured and Unstructured Data*

85. The answer is: Correlation

    If you are unsure why, please review *Chapter 10, Introduction to Inferential Statistics – Calculating Correlations*

86. The answer is: Data encryption

   If you are unsure why, please review *Chapter 14, Data Governance – Understanding Data Security*

87. The answer is: T-test

   If you are unsure why, please review *Chapter 10, Introduction to Inferential Statistics – Understanding T-Tests*

88. The answer is: Streamlining data access

   If you are unsure why, please review *Chapter 15, Data Quality and Management – Understanding Master Data Management (MDM)*

89. The answer is: Subscription

   If you are unsure why, please review *Chapter 12, Reporting Process – Understanding Report Delivery*

90. The answer is: Data deletion

   If you are unsure why, please review *Chapter 14, Data Governance – Knowing Use Requirements*

Index

‹packt›

Other Books You May Enjoy

If you enjoyed this book, you may be interested in these other books by Packt:

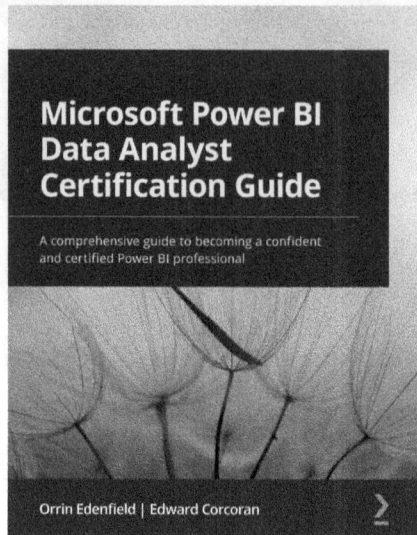

Microsoft Power BI Data Analyst Certification Guide

Orrin Edenfield, Edward Corcoran

ISBN: 978-1-80323-856-2

- Connect to and prepare data from a variety of sources
- Clean, transform, and shape your data for analysis
- Create data models that enable insight creation
- Analyze data using Microsoft Power BI's capabilities
- Create visualizations to make analysis easier
- Discover how to deploy and manage Microsoft Power BI assets

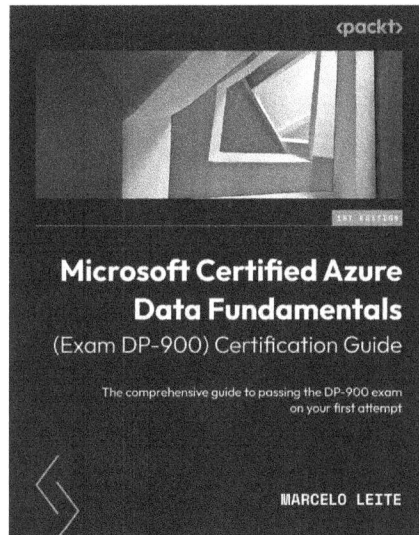

Microsoft Certified Azure Data Fundamentals (Exam DP-900) Certification Guide

Marcelo Leite

ISBN: 978-1-80324-063-3

- Explore the concepts of IaaS and PaaS database services on Azure
- Query, insert, update, and delete relational data using SQL
- Explore the concepts of data warehouses in Azure
- Perform data analytics with an Azure Synapse Analytics workspace
- Upload and retrieve data in Azure Cosmos DB and Azure HDInsight
- Provision and deploy non-relational data services in Azure
- Contextualize the knowledge with real-life use cases
- Test your progress with a mock exam

Packt is searching for authors like you

If you're interested in becoming an author for Packt, please visit `authors.packtpub.com` and apply today. We have worked with thousands of developers and tech professionals, just like you, to help them share their insight with the global tech community. You can make a general application, apply for a specific hot topic that we are recruiting an author for, or submit your own idea.

Share Your Thoughts

Now you've finished *CompTIA Data +: DAO-001 Certification Guide*, we'd love to hear your thoughts! Scan the QR code below to go straight to the Amazon review page for this book and share your feedback or leave a review on the site that you purchased it from.

`https://packt.link/r/1-804-61608-7`

Your review is important to us and the tech community and will help us make sure we're delivering excellent quality content.

Download a free PDF copy of this book

Thanks for purchasing this book!

Do you like to read on the go but are unable to carry your print books everywhere? Is your eBook purchase not compatible with the device of your choice?

Don't worry, now with every Packt book you get a DRM-free PDF version of that book at no cost.

Read anywhere, any place, on any device. Search, copy, and paste code from your favorite technical books directly into your application.

The perks don't stop there, you can get exclusive access to discounts, newsletters, and great free content in your inbox daily

Follow these simple steps to get the benefits:

1. Scan the QR code or visit the link below

https://packt.link/free-ebook/9781804616086

2. Submit your proof of purchase
3. That's it! We'll send your free PDF and other benefits to your email directly

www.ingramcontent.com/pod-product-compliance
Lightning Source LLC
Chambersburg PA
CBHW082132210326
41599CB00031B/5956